General View of the Agriculture of the County of Lancaster

With Observations on the Means of Its Improvement

JOHN HOLT

CAMBRIDGE
UNIVERSITY PRESS

CAMBRIDGE
UNIVERSITY PRESS

University Printing House, Cambridge, CB2 8BS, United Kingdom

Cambridge University Press is part of the University of Cambridge.
It furthers the University's mission by disseminating knowledge in the pursuit of
education, learning and research at the highest international levels of excellence.

www.cambridge.org
Information on this title: www.cambridge.org/9781108083263

© in this compilation Cambridge University Press 2018

This edition first published 1795
This digitally printed version 2018

ISBN 978-1-108-08326-3 Paperback

This book reproduces the text of the original edition. The content and language reflect
the beliefs, practices and terminology of their time, and have not been updated.

Cambridge University Press wishes to make clear that the book, unless originally published
by Cambridge, is not being republished by, in association or collaboration with,
or with the endorsement or approval of, the original publisher or its successors in title.

The original edition of this book contains a number of oversize plates
which it has not been possible to reproduce to scale in this edition.
They can be found online at www.cambridge.org/9781108083263

CAMBRIDGE LIBRARY COLLECTION

Books of enduring scholarly value

Botany and Horticulture

Until the nineteenth century, the investigation of natural phenomena, plants and animals was considered either the preserve of elite scholars or a pastime for the leisured upper classes. As increasing academic rigour and systematisation was brought to the study of 'natural history', its subdisciplines were adopted into university curricula, and learned societies (such as the Royal Horticultural Society, founded in 1804) were established to support research in these areas. A related development was strong enthusiasm for exotic garden plants, which resulted in plant collecting expeditions to every corner of the globe, sometimes with tragic consequences. This series includes accounts of some of those expeditions, detailed reference works on the flora of different regions, and practical advice for amateur and professional gardeners.

General View of the Agriculture of the County of Lancaster

The author and schoolteacher John Holt (1743–1801), who lived near Liverpool, wrote novels, historical works, and many contributions to the Gentleman's Magazine, including a monthly meteorological report. This survey of agriculture in Lancashire was published in 1794, with an enlarged edition (reissued here) following in 1795. The book begins with the detailed plan of the Board of Agriculture for obtaining regional surveys, such as the 'Rural Economies' of William Marshall (several of which are reissued in this series). It goes on to describe the geography, climate and soils of Lancashire, its land holdings and land use, noting the manufacturing as well as the agricultural areas (and the improvement to the latters' soil by the spreading of 'night-soil' from the growing cities). A detailed examination of the livestock, agrarian and horticultural products of the county follows, with figures on farming costs and crop yields. This remains a fascinating resource for social and agricultural historians.

Cambridge University Press has long been a pioneer in the reissuing of out-of-print titles from its own backlist, producing digital reprints of books that are still sought after by scholars and students but could not be reprinted economically using traditional technology. The Cambridge Library Collection extends this activity to a wider range of books which are still of importance to researchers and professionals, either for the source material they contain, or as landmarks in the history of their academic discipline.

Drawing from the world-renowned collections in the Cambridge University Library and other partner libraries, and guided by the advice of experts in each subject area, Cambridge University Press is using state-of-the-art scanning machines in its own Printing House to capture the content of each book selected for inclusion. The files are processed to give a consistently clear, crisp image, and the books finished to the high quality standard for which the Press is recognised around the world. The latest print-on-demand technology ensures that the books will remain available indefinitely, and that orders for single or multiple copies can quickly be supplied.

The Cambridge Library Collection brings back to life books of enduring scholarly value (including out-of-copyright works originally issued by other publishers) across a wide range of disciplines in the humanities and social sciences and in science and technology.

The material originally positioned here is too large for reproduction in this reissue. A PDF can be downloaded from the web address given on page iv of this book, by clicking on 'Resources Available'.

GENERAL VIEW

OF THE

AGRICULTURE

OF THE COUNTY OF

LANCASTER:

WITH OBSERVATIONS ON THE MEANS OF ITS IMPROVEMENT.

Drawn up for the Confideration of the

BOARD OF AGRICULTURE AND INTERNAL IMPROVEMENT,

From the Communications of Mr. *JOHN HOLT,*
of WALTON, near LIVERPOOL;

And the additional Remarks of feveral refpectable GENTLEMEN and
FARMERS in the County.

———————————

Prima Ceres ferro mortales vertere terram
Inftituit ————
Dicendum eft, quæ fint duris agreftibus arma,
Queis finè, nec potuére feri, nec furgere meffes.

GEORGICA.

See the fun gleams; the living paftures rife,
After the nurture of the fallen fhower,
How beautiful! How blue the ethereal vault,
How verdurous the lawns, how clear the brooks!
Such noble warlike fteeds, fuch herds of kine,
So fleek, fo vaft; fuch fpacious flocks of fheep,
Like flakes of gold, illumining the green,
What other paradife adorn but thine,
Britannia? Happy, if thy fons would know
Their happinefs. To thefe thy naval ftreams,
Thy frequent towns fuperb of bufy trade,
And ports magnific add, and ftately fhips
Innumerous. ————

DYER.

═══════════════════

LONDON:

Printed for G. NICOL, PALL-MALL,
Bookfeller to HIS MAJESTY, and to the BOARD of AGRICULTURE;

And fold by Meffrs. ROBINSON, Paternofter-Row; J. SEWELL, Cornhill;
CADELL and DAVIES, Strand; WILLIAM CREECH, Edinburgh;
and JOHN ARCHER, Dublin. 1795.

ADVERTISEMENT.

THE great defire that has been very generally expreffed, for having the AGRICULTURAL SUR-VEYS of the KINGDOM re-printed, with the additional Communications which have been received fince the ORIGINAL REPORTS were circulated, has induced the BOARD OF AGRICULTURE, to come to a Refolution, of re printing fuch as may appear on the whole fit for Publication. It is proper at the fame Time to add, that the Board does not confider itfelf refponfible, for any Fact or Obfervation contained in the Reports thus re-printed, as it is impoffible to confider them yet in a perfect State; and that it will thankfully ac-knowledge any additional Information which may ftill be communicated: An Invitation, of which, it is hoped, many will avail themfelves, as there is no Circumftance from which any one can derive more real Satisfaction, than that of contributing, by every poffible means, to promote the Improvement of his Country.

N. B.—*Letters to the Board, may be addreffed to Sir* JOHN SINCLAIR, Bart. *the Prefident,* M. P. *London.*

LONDON, June 1795.

P L A N

FOR RE-PRINTING THE

Agricultural Surveys.

By the PRESIDENT of the Board of Agriculture.

A BOARD eftablifhed for the purpofe of making every
effential enquiry, into the Agricultural State, and the
means of promoting the internal improvement of a powerful
Empire, will neceffarily have it in view, to examine the fources
of public profperity, in regard to various important particulars.
Perhaps the following is the moft natural order for carrying on
fuch important inveftigations; namely, to afcertain,

1. The riches to be obtained from the furface of the national
   territory.
2. The mineral or fubterraneous treafures of which the
   country is poffeffed.
3. The wealth to be derived from its ftreams, rivers, canals, inland navigations, coafts, and fifheries : And
4. The means of promoting the improvement of the people
   in regard to their health, induftry, and morals, founded
   on a *ftatiftical* furvey, or a minute and careful enquiry
   into the actual ftate of every parochial diftrict in the
   kingdom, and the circumftances of its inhabitants.

b   Under

Under one or other of thefe heads, every point of real importance, that can tend to promote the general happinefs of a great
nation, feems to be included.

Inveftigations of fo extenfive and fo complicated a nature,
muft require, it is evident, a confiderable fpace of time before
they can be completed. Differing indeed in many refpects
from each other, it is better perhaps that they fhould be undertaken at different periods, and feparately confidered. Under
that impreffion, the Board of Agriculture has hitherto directed
its attention to the firft point only, namely the cultivation of
the furface, and the refources to be derived from it.

That the facts effential for fuch an inveftigation, might be
collected with more celerity and advantage, a number of intelligent and refpectable individuals were appointed, to furnifh the
Board with accounts of the ftate of hufbandry, and the means
of improving the different diftricts of the kingdom. The returns they fent were printed, and circulated by every means the
Board of Agriculture could devife, in the diftricts to which they
refpectively related; and in confequence of that circulation, a
great mafs of additional valuable information has been obtained. For the purpofe of communicating that information
to the Public in general, but more efpecially to thofe counties
the moft interefted therein, the Board has refolved to re-print
the Survey of each County, as foon as it feemed to be fit for
publication; and among feveral equally advanced, the counties
of Norfolk and Lancafter were pitched upon for the commencement of the propofed publication; it being thought moft
advifable, to begin with one county on the Eaftern, and another on the Weftern coaft of the ifland. When all thefe Sur-
veys fhall have been thus re-printed, it will be attended with little
difficulty to draw up an abftract of the whole (which will not
probably exceed two or three volumes quarto) to be laid be-

*
                                                                fore

fore His Majefty, and both Houfes of Parliament; and afterwards, a general Report on the prefent ftate of the country and
the means of its improvement, may be fyftematically arranged,
according to the various fubjects connected with agriculture.
Thus every individual in the kingdom may have,

1. An account of the hufbandry of his own particular
county; or,

2. A general view of the agricultural ftate of the kingdom
at large, according to the counties, or diftricts, into
which it is divided; or,

3. An arranged fyftem of information on agricultural fubjects, whether accumulated by the Board fince its
eftablifhment, or previoufly known;

And thus information refpecting the ftate of the kingdom, and
Agricultural knowledge in general, will be attainable with
every poffible advantage.

In re-printing thefe Reports, it was judged neceffary, that they
fhould be drawn up according to one uniform model; and after
fully confidering the fubject, the following form was pitched
upon, as one that would include in it all the particulars which
it was neceffary to notice in an Agricultural Survey. As
the other Reports will be re-printed in the fame manner,
the reader will thus be enabled to find out at once, where any
point is treated of, to which he may wifh to direct his attention.

PLAN

PLAN OF THE RE-PRINTED REPORTS.

Chap.

---

* Where the quantity is confiderable, the information refpecting the crops commonly cultivated, may be arranged under the following heads :

1. Preparation { tillage, manure. }

2. Sort.
3. Steeping.
4. Seed (quantity fown.)
5. Time of fowing.

6. Culture whilft growing { hoe, weeding feeding. }

7. Harveft.
8. Threfhing.
9. Produce.
10. Manufacture of bread.

In general the fame heads will fuit the following grains :
Barley. Oats. Beans. Rye. Peafe. Buck-wheat.

Vetches - - - Application.

Cole-feed - { Feeding, Seed. }

Turneps - - { Drawn - - - - - - Fed - - - - - - - - Kept on grafs - - - —— in houfes - - - }

XIII. Live

PERFECTION

PERFECTION in fuch inquiries is not in the power of any body of men to obtain at once, whatever may be the extent of their views, or the vigour of their exertions. If Lewis XIV. eager to have his kingdom known, and poffeffed of boundlefs power to effect it, failed fo much in the attempt, that of all the provinces in his kingdom, only one was fo defcribed as to fecure the approbation of pofterity * ; it will not be thought ftrange that a Board, poffeffed of means fo extremely limited, fhould find it difficult to reach even that degree of perfection which, perhaps, might have been attainable with more extenfive powers. The candid Reader cannot expect in thefe Reports more than a certain portion of ufeful information, fo arranged as to render them a bafis for further and more detailed enquiries. The attention of the intelligent Cultivators of the kingdom, however,

* See Voltaire's Age of Lewis XIV. vol. ii. p. 127, 128, edit. 1752.

The following extract from that work will explain the circumftance above alluded to.

" Lewis had no Colbert, nor Loùvois, when about the year 1698, for
" the inftruction of the Duke of Burgundy, he ordered each of the inten-
" dants to draw up a particular defcription of his province. By this means
" an exact account of the kingdom might have been obtained, and a
" juft enumeration of the inhabitants. It was an ufeful work, though all
" the intendants had not the capacity and attention of Monfieur de La-
" moignon de Baville. Had what the king directed been as well executed
" in regard to every province, as it was by this magiftrate in the account
" of Languedoc, the collection would have been one of the moft valuable
" monuments of the age. Some of them are well done ; but the plan was
" irregular and imperfect, becaufe all the intendants were not reftrained
" to one and the fame. It were to be wifhed, that each of them had given,
" in columns, the number of inhabitants in each election ; the nobles, the
" citizens, the labourers, the artifans, the mechanics, the cattle of every
" kind ; the good, the indifferent, and the bad lands, all the clergy, regu-
" lar and fecular, their revenues, thofe of the towns, and thofe of the
" communities.

" All thefe heads, in moft of their accounts, are confufed and imper-
" fect ; and it is frequently neceffary to fearch with great care and pains
" to find what is wanted. The defign was excellent, and would have been
" of the greateft ufe, had it been executed with judgment and unifor-
" mity."

will

will doubtlefs be excited, and the minds of men in general gradually brought to confider favourably of an undertaking, which will enable all to contribute to the national ftores of knowledge, upon topics fo truly intereſting as thofe which concern the Agricultural interefts of their country; interefts, which on juft principles never can be improved, until the prefent ftate of the kingdom is fully known, and the means of its future improvement afcertained with minutenefs and accuracy.

PRELIMINARY OBSERVATIONS

TO THE

LANCASHIRE RE-PRINTED REPORT.

———————

IN the courfe of an addrefs to the. Board of Agriculture, when it firft affembled, on the 4th of September 1793, I took an opportunity of ftating the meafures which feemed to me the moft likely to promote the objefts of that inftitution; and fubmitted to the confideration of the Board, whether the firft objeft ought not to be, *to afcertain facts*, without which no theory or fyftem of reafoning, however plaufible, could be depended on; that for attaining fo important an objeft, it would be neceffary to examine into the agricultural ftate of all the different counties in the kingdom, and to enquire into the means, which, in the opinion of intelligent men, were the moft likely to promote either a general fyftem of improvement, or the advantage of particular diftrifts; that by employing a number of able men for that purpofe, by circulating their reports previous to their being publifhed, and by requefting the additional remarks and obfervations of thofe to whom fuch communications were fent, it was probable that no important faft, or even ufeful idea, would efcape notice.

The plan thus chalked out having been approved of by the Board, it was immediately fet about with every poffible degree

of

of energy. Among other intelligent individuals nominated for that purpose, Mr. Holt, of Walton, near Liverpool, was appointed to take a furvey of the county of Lancafter. Thofe who have had an opportunity of examining his original Report, will fee the pains which he took to fulfil the objects of his miffion. As foon as his Report was printed, it was circulated throughout the county, for the purpofe of obtaining additional information; and though, from the want of the privilege of fending and receiving packets duty-free, (a privilege which, it is hoped, the Board will foon obtain, for the want of it impedes all its operations) the circulation of the Report was attended with confiderable difficulty and expence; yet, on the whole, fuch a number of copies were returned, with valuable additional obfervations, as to induce the Board to form an opinion, that the work might now be rendered fit for publication; and that it would be defirable to take the fenfe of the Public refpecting the beft mode of communicating the information which it had thus accumulated, by re-printing the corrected accounts of two counties, namely, Norfolk on the Eaftern, and Lancafhire on the Weftern coaft of the ifland.

There is every reafon to believe, that the accounts of the other diftricts in the kingdom will foon be equally complete; in which cafe, a greater mafs of agricultural knowledge will be collected, in a fpace of little more than two years, than probably can be found in all preceding publications on the fubject of hufbandry; and thus the foundation will be laid for a general fyftem of improvement, on that beft and fureft of all foundations, a knowledge of facts.

Next

Next to collecting information, the improvement of a country muſt depend upon rouſing a proper ſpirit of exertion, in order that the information thus accumulated may be put into action. By the happy conſtitution of this country, and the wiſdom of its laws, property is better ſecured here than perhaps in any other ſtate that ever exiſted, which undoubtedly is a great ſpur to exertion. But the legiſlature ſeems to have truſted too much to the beneficial effects of that ſecurity, and to think that no other encouragement or ſpur could be neceſſary. Fortunately, however, a new ſyſtem has commenced. Parliament has already begun to vote ſome aid for the improvement of huſbandry. The legiſlature has at laſt taken the plough and the ſpade under its immediate protection; and thoſe who make any uſeful diſcovery likely to be of ſervice to Agriculture, have now every reaſon to expect attention to their claims, and that encouragement which their diſcoveries may be found to merit.——As ſome of the principal improvements which Mr. Elkington, the celebrated drainer, to whom Parliament has lately granted 1000 *l.* were made in Lancaſhire, it was natural to allude to that circumſtance, when the Report of that county came under conſideration.

Where both ſkill in agriculture, and a ſpirit of improvement, exiſt, there can be but one thing wanting, namely, capital. There is little riſk, however, of any deficiency of that nature in theſe kingdoms, unleſs our capital ſhould be diverted from its natural means of employment, *domeſtic improvement,* to remote and foreign ſpeculations. The beſt mode of preventing ſuch a deviation ſeems to be, *to make the*

2   *kingdom*

kingdom known to its inhabitants, and to point out the benefit which they may derive from improving it. Such are the objects of thefe inquiries; which, fo far as they concern the county of Lancafter, feem to have made very confiderable progrefs, although in fome particulars they have not reached that degree of perfection that would be fo truly defirable; but which probably will yet be attained, even previous to the conclufion of the prefent century, when the Statiftical Account of Great Britain, that moft important of all the labours which the Board of Agriculture can undertake, is completed.

AGRICULTURAL SURVEY

OF

LANCASHIRE.

CHAPTER I.

GEOGRAPHICAL STATE AND CIRCUMSTANCES.

SECT. I.—*Situation and Extent.*

L ANCASHIRE is a maritime county, bounded on the coaſt by Saint George's channel and the Iriſh ſea.

The dimenſions of the county are as follows *.—Its greateſt length 74 miles; breadth 44½ miles.—Its circumference (croſſing the Ribble, at Heſketh bank) 342 miles; containing 1,765 ſquare miles, and 1,129,600 ſtatute acres.— Total number of pariſhes, with the additional ones, 62.

SECT. 2.—*Diviſions.*

THE county is divided into ſix hundreds; namely, Salford, Weſt Derby, Leyland, Blackburne, Amoundernefs, and Lonſdale. There are two diſtricts in it which may deſerve more particular mention; namely, the Filde, which is remarkable

---

* Calculated upon this occaſion by Mr. William Yates, who ſurveyed and publiſhed a map of the county of Lancaſter in the year 1786.

for its great fertility; and Furnefs, bordering on Cumberland and Weftmorland, where there is a fertile vale. The
Filde is peculiarly diftinguifhed for its breed of cattle.
Since the circulation of the Lancafhire Report there, a new
fpirit for agricultural improvements has arifen, particularly in regard to draining, watering, making compofts, manuring their lands, &c. which cannot fail to be attended
with the beft confequences.

The fhape of the county is fomewhat fimilar to that of
England, Wales, and part of Scotland; *e. g.* fuppofe the parts
beyond the fands reprefent part of Scotland; the river Loyne,
and the inlet which runs up to Cockerham, the rivers Merfey
and Dee; that tract called the Filde, the principality of Wales;
the Ribble, Briftol Channel, and the Severn; and, again, the
river Merfey, the fouthern boundary of the county, by the
Englifh Channel, the fouthern boundary of the kingdom. The
indentures upon the eaftern parts of the county have a ftrong
fimilarity to the indentures on the eaftern part of the kingdom.

Sect. 3.—*Climate.*

THE ridge of mountains, which bounds this county on
the eaftern fide, from Yorkfhire, and which runs not only
through Yorkfhire, but Chefhire, Derbyfhire, and Stafford-
fhire, &c. and called, not improperly, the Back-bone of the
kingdom, being the moft elevated ground in the ifland, fcreens
Lancafhire more particularly from the ungenial eaftern blafts,
the frofts, blights, and infects, which infeft the countries bordering upon the German ocean; and though the high mountains may caufe a greater quantity of rain to fall in this diftrict,
(as appears by rain-gauges kept for that purpofe) than in the
more interior parts of the kingdom; yet this county, fanned
with the weftern gales, or north-weft breezes, has a falubrity
of air, to which may be attributed the vigour and activity of the
inhabitants, who are, if temperate, generally long-lived. The
faline particles, with which the wefterly winds are loaded, may
alfo not a little contribute to the verdure of the fields. Snow

continues

continues but a short space of time upon the ground, owing to the maritime situation of the county.

The prevailing winds of this county are the West and N. W. winds, which produce a mildness of climate, and salubrity of air and atmosphere, unknown in most districts so far advanced to the north.

Though that part of the county which lies to the south of the river Ribble is in general a low and flat region, perhaps few districts of this or any other kingdom can produce a more healthy, vigorous, or active race of inhabitants; living in general, when temperate, to a great age, and bearing in the whole of their appearance a most ample testimony, to the salubrity of their native air. The beauty of the *Lancashire witches* has long been celebrated; and the men are no less distinguished for their military strength and prowess. The neighbourhood of the Atlantic ocean, and the elevation of its mountain boundary, certainly render this county more subject to wet weather than most in the kingdom. These frequent rains, however, have the effect of rendering Lancashire one of the most productive and certain grass-land districts in the island. The soil is peculiarly adapted to grass, and the climate uncommonly favourable for that production.

Perpendicular

Perpendicular Height of the RAIN that has fallen at Lancaster, during the last Nine Years; distinguishing each Month and Year in Incnes and Lines. By DR. CAMPBELL, of Lancaster.

Years - -	1784.	1785.	1786.	1787.	1788.	1789.	1790.	1791.	1792.
Months.	In. Lin.	In. Lin.	In. Lin.	In. Lin	In. Lin.	In. Lin.	In. Lin.	In. Lin.	In. Lin.
January - -	2 8¼	2 6	2 6	1 7¼	2 10	4 5	5 11	5 10	3 2
February -	2 3¼	– 6⅓	1 1	5 –	2 1	4 11	1 2¼	3 1¼	3 –
March - -	2 7½	– 1	– 11	3 7⅓	1 10	– 8⅓	– 8	2 2	5 9
April - -	3 –	1 8	– 4½	1 3½	2 7½	4 3¼	1 3½	4 3	5 9½
May - -	3 –	1 6	1 8	1 4¼	1 1	4 1½	2 1	2 4½	5 –
June - -	5 9	1 –½	1 10	3 6¼	2 –⅓	5 2½	4 –	– 10⅓	3 10
July - -	3 –	2 1½	2 9½	7 –	6 5	5 7¼	7 6	3 6	5 1¼
August - -	5 –	10 4	5 –	7 –	2 –¼	– 5½	3 10⅓	6 2	8 6
September -	2 7	5 6	7 11	2 –	3 7¼	4 1¼	5 5	1 9½	9 4
October - -	– 8	5 9	1 6	9 9	2 1½	6 6	2 9	3 10	4 3
November -	3 –	4 6	3 –	4 5¼	1 9¼	4 1	4 6⅓	6 6	4 –
December -	1 6	1 2	3 8	4 5	– 10⅓	6 6	7 4	5 7½	8 1
Total -	35 1½	36 9⅓	32 3	51 –¼	29 4½	51 –¼	46 6½	46 –¼	65 10

N. B.—A line is the twelfth part of an inch.

Mean

Mean heat of the Thermometer at noon at Lancaster - - 51.8.
                    D° - - at London - - 56.
                    D° - - at Edinburgh - 50.1.

WINDS blow at Lancaster:

North - - 30 Days.
N. E. - - 67
S. E. - - 35
East - - 17
South - - 51
S. W. - - 92
N. W. - - 26
West - - 47

The mean heat of the Thermometer, and the direction of the Winds, are taken from an average of the seven years from 1784 to 1790 inclusive.

> Perpendicular Height of R A I N that has fallen at Liverpool from the year 1784 to the year 1792 inclusive. By MR. WILLIAM HUTCHINSON, late Dock-master.

1784.	1785.	1786.	1787.	1788.	1789.	1790.	1791.	1792.
36¼ In.	26¼ In.	26½ In.	37⅓ In.	24⅛ In.	48¼ In.	42¼ In.	45⅝ In.	54¼ In.

The feed-time, and harvest, vary a little between the northern and southern parts of the county. Those towards the east, and contiguous to the mountains, are in general later than the south-western parts.

The following Register will shew, that there is a greater difference of season than many may imagine; and if these meteorological registers were multiplied, and kept in different places, and the system more extended, such data would not only be pleasing memoranda, but afford many useful hints.

The following particulars were taken from the memoranda of D. Daulby, Esq. Birch House, Liverpool, respecting some articles produced on the grounds of Mr. Hill, of Wallasey, in Cheshire, about three miles from Liverpool. The articles mentioned were for the Liverpool market, the dates
corresponding

corresponding to the two days in the week on which the market is held, Wednesday or Saturday. It may be worthy of remark, that there is a general strife betwixt the Kirkdale and Wallasey gardeners, who can produce the first early potatoe at Liverpool market. They generally succeed both on the same day. In the year 1790 the Cheshire gardener had, however, the start by nearly a whole week.

EARLY POTATOES.

1766. June 7,   20 lb. sold for 5 d. and 6 d. per lb.
1767. June 6,   3 lb. sold for 14 d. in the whole.
1768. May 14,   8 lb. sold for 4 s. 8 d.
1769. May 13,   2 lb. sold for 1 s,
1770. May 23,   2 lb. for 3 s.
1771. May 18,   ½ lb. for 1 s.
1772. May 13,   1 lb. for 2 s. 6 d.

N. B.—From this period the early potatoes have been regularly sold for 2 s. 6 d. per lb. when first brought to market.

After this period the Register was extended to the following articles; namely,

	ASPARAGUS.	POTATOES.	GOOSEBERRIES.
1773.	April 10.	April 7.	May 5.
1774.	3.	30.	9.
1775.	1.	19.	April 26.
1776.	6.	17.	May 2.
1777.	4.	24.	12.
1778.	11.	25.	9.
1779.	March 27.	3.	April 10.
1780.	April 15.	20.	May 6.
1781.	March 31.	14.	April 21.
1782.	May 4.	May 11.	May 18.
1783.	April 12.	1.	April 30.
1784.	May 8.	17.	May 22.
1785.	April 23.	14.	18,
1786.	22.	13.	10.
1787.	March 28.	April 11.	April 28.

1788,

1788.	April 19.	May 11.	May 7.
1789.	18.	9.	9.
1790.	3.	April 3.	April 24.
1791.	9.	16.	23.
1792.	7.	25.	25.
1793.	May 1.	May 11.	May 18.
1794.	April 15.	April 12.	April 18.

From the above Regifter it appears, that the difference between an early and late fpring is not lefs than fix wecks; *e. g.*

	ASPARAGUS.	POTATOES.	GOOSEBERRIES.
1789.	March 27.	April 3.	April 10.
1784.	May 8.	May 17.	May 22.

From this Regifter may alfo be traced, the improved cultivation of the early potatoe upon common ground: but the potatoe at prefent may be truly faid to be raifed the whole year throughout, by the new method of heating the ftoves with fteam. Mr. Butler, gardener to the Earl of Derby, at his feat at Knowfley, has practifed this fome time; and Mr. Collins, late his lordfhip's gardener, who has ground near Liverpool, had, under glaffes, forced by the heat of fteam, Chriftmas, 1794, nearly, as he calculated, one cwt. of potatoes, ready to take up. But he obferved, that the procefs by fteam was too expenfive to afford any profit at the price they were ufually fold.

It will at this day fcarcely be credited, that when potatoes began to be brought to market fo early as June, the gardeners were under the neceffity of bringing the ftems adhering to the potatoes, for without this no purchafer could be obtained.

A gentleman who has been particularly attentive to this fubject, obferved that, in this northern diftrict, autumnal feeds require to be committed to the earth one fortnight at leaft earlier than is recommended by Mawe, in his Kalendar.

S E C T. 4.—*Soil and Surface.*

THE features of this county are, in many places, ſtrongly marked; towards the north they are bold and picturefque, diverſified with alpine mountains and fertile vales. The north-eaſt part of the county, Blackburn, Clithero, Haſlingden, &c. is rugged, interſperſed with many rivulets, with a thin ſtratum of upper ſoil; the ſouthern part more ſoftened, and the plains are more fertilized: along the ſea coaſt, the land is chiefly flat, and has the appearance, in many places, as if formerly covered by the ocean. In various fields at Formby, near the ſhore, there is ſoil above two feet below the ſand, which lies beneath the preſent green-ſward. There are the ſtrongeſt reaſons for believing that this ſoil (which is about four inches thick) originally formed the ſurface of the ground, and was gradually buried by ſand from the neighbouring hills. Few countries produce greater varieties of ſoil, which yet does not change ſo rapidly as in ſome others.

The greateſt proportion of that diſtrict, which lies between the Ribble and the Merſey, has for its ſuperficies a ſandy loam, well adapted to the production of almoſt every vegetable that has yet been brought under cultivation, and that to a degree which renders it impoſſible to eſtimate the advantage which might be obtained, by improved and ſuperior management. The ſubſtratum of this ſoil is generally the red rock, or clay-marle, an admirable ſandy loam, perhaps one of the moſt deſirable ſoils that can be found, equally well adapted to the production of every vegetable. In this diſtrict there is little or no gravelly ſoil, no chalk or flint, no ſtony land, and very little obdurate clay, for the generality of it (except what is under graſs, and indeed much of that) is treated in a manner that does little credit to this æra of improved and enlightened agriculture. There is alſo a black ſandy loam, ſomething diſtinct from the above deſcription, which has no red rock, but the ſubſtratum white ſand, under which is clay, and then marle. There are alſo tracts of white ſand lands, and ſome little pebbly-gravel lands. There are many large tracts which

come

come under the denomination of *moſſes*, and ſome ſtiff, but no obdurate clay lands.

There is a kind of land which throws up great quantities of ruſhes, not owing ſo much to ſprings, as to a thin ſtratum of ſurface ſoil, under which lies a bed of matter, principally compoſed of clay, which does not admit the water to penetrate ; therefore, the upper ſurface, or ſoil, is kept in a continual ſpongy ſtate (if not ſurface-drained) and produces ruſhes and other ſour graſſes *.

Remarks on this Obſervation †.—" This kind of land lying upon clay or marle, is not (I am of opinion) cured, and but little benefitted, by ſurface-draining ; the evil, generally, if not always, is under the ſoil, occaſioned by the ‡ ſand-beds, which are of various depths and forms, from the ſurface of the clay or marle, ſay from half a yard to $1\frac{1}{2}$ yard deep, like ſo many baſons filled with ſand and water, which keep the ſoil continually moiſt (except in very dry weather) ; and as it muſt be granted by all, that ruſhes are occaſioned by a ſtagnated moiſture, ſo it is obvious, that the only effectual method of cure is draining the land ſufficiently deep, ſo that the wet cannot be ſucked up to the ſoil or upper ſurface. If land lies wet in winter, when dried by ſummer's heat it becomes hard and firmly baked together, ſo that little vegetation is produced, conſequently the propriety of under-draining, to produce crops and deſtroy ruſhes, is obvious. And further, to prove this, if a drain is cut acroſs a field of this deſcription into the marle, and through the ſand-beds, it will be found, that there is a continual ſtream all the winter ſeaſon.

---

* Upon ſuch land, common rack or gravel ſand is ſpread upon the ground previous to ploughing, as thick as common dunging ; and then, after ſecond ploughing or croſs-cutting, dung at top, and harrow in the ſeed, and you will looſen the water-tight ſoil. If twice repeated, the ſucceſs is infallible.

† By Mr. James Blundell.

‡ Called here ſand-goats.

C

" Good

" Good marle has the property of ftiffening light land, and
meliorating, and unbinding, (if dry) ftiff land.   Stiff foil, it is
true, for a long time, refifts the rain before it is faturated;
but when made wet, it longer retains the moifture than a dry
foil."

Marle has a good effect upon thefe lands; for, befides its
ufual qualities of promoting putrefaction, it renders the foil
ftiffer, and enables it to refift and throw off the furface water
more effectually.

Moor lands which are in a ftate of nature, and produce heath,
and other wild plants, are of various qualities ; very extenfive
indeed, and much more fo than might have been expected in a
county fo populous, and confequently where lands muft be fo
valuable.

Thefe are diftinctions, not neceffary perhaps, on this occafion, to particularize more minutely, than by obferving that the
vales are in general fertile, but have lefs of that fertility as
they approach nearer to the higher lands.

S e c t. 5.—*Water.*

The great advantages which this county poffeffes, both
from its having fuch a range of fea-coaft, and alfo from the
numerous ftreams and rivers it is poffeffed of (not forgetting
the lakes of Windermere and Conifton-water) need hardly
be dwelt upon, being fo extremely obvious.   It may be fufficient to remark, that without thofe advantages, neither the manufactures of the county, nor the fea and inland fifheries,
a matter of no inconfiderable moment to the inhabitants, could
be carried on to the fame extent.

It is believed, that the only decoy pond is at Orford, the
feat of John Blackburne, Efq. member for the county.

S e c t.

Sect. 6.—*Minerals.*

LANCASHIRE has fome local advantages, which have been the caufe of rendering the county fo famous for its manufactures. Thefe in a great meafure depend upon two moft material articles, coals and water; the former of which lie in immenfe beds towards the fouthern and middle part, and the various rivers, &c. which, together with the fprings, in fo many places interfect the county, have conjointly had no fmall effect upon the agriculture of this diftrict, as will be feen hereafter. The north and north-eaft diftricts produce lime-ftone in abundance, but no calcareous matter except marle is found towards the fouth; a fmall quantity of lime-ftone pebble upon the banks of the river Merfey is alfo to be excepted. In the townfhip of Halewood, near Liverpool, lime-ftone is found and got at different depths, but in fmall quantities.

Coals have not been found, as it is faid, farther north in the county than Chorley and Colne. The next bed of that ufeful article, after a long fpace, appears again in immenfe quantities at Whitehaven and Newcaftle-upon-Tyne. The cannel (a fpecies of coal refembling black marble) lies chiefly at Haigh, near Wigan, and occupies a fpace, as it is faid, of about four miles fquare.

Near Leigh is found lime of a peculiar quality, which refifts the effects of water, and is therefore applied to the conftruction of cifterns to hold water, and mortar for building under water. Alfo at Ardwick, near Manchefter, not many years ago has been difcovered a lime of fimilar, by fome it is faid of fuperior, quality. The tarras-ciftern at Drury-lane houfe is of this lime.

There is faid to be coal about Hornby.—" The exportation of coal to foreign countries is now become a great trade, which has encreafed rapidly within thefe few years paft. No doubt this trade is beneficial to fome individuals, and alfo to the revenue; at the fame time we fhould reflect, that this avidity for prefent profit has a deadly fting in the tail of it. We are now eagerly fupplying other nations, and frequently our worft enemies, with thofe coals, which the metropolis and our manufac-

　　　　　　　　　tories

tories will one day ftand in need of; and even now, the exportation trade has raifed the price of coals fo high, as to be exceedingly inconvenient to many of our manufactories, and to
many of the induftrious poor, efpecially in South Britain; and
this in lefs than one hundred years from the commencement of
an extenfive confumption.   What may we fuppofe will be the
price of them two hundred years hence * ? "

To the above may be added, the great confumption of coals
by fire-engines, many of which, in the courfe of years, muft
ftand ftill for want of that article.

The power of fteam is fo great, and the mechanifm of the engine is fo much improved, that in a little time it is probable
we may fee a fmall veffel over the kitchen fire with hot water,
forcing an engine and working the churn.

Befides coal, this county alfo produces ftone, of various
denominations.   Near Lancafter (upon the common) is an
extenfive quarry of excellent free-ftone, which admits of a
fine polifh.   The county town (Lancafter) is built wholly of
this ftone, and, for its neatnefs, is excelled by few towns in the
kingdom.   Flaggs and grey flates are dug up at Holland, near
Wigan.   Blue flates are got in large quantities in the mountains, called Coniftone and Telberthwaite fells, near Hawks-
head, of which many are exported.   They are chiefly divided
into three claffes, viz. London, country, and *tom* flate, which
are valued in a due proportion: London are the beft, &c.
The beft fcythe-ftones are obtained at Rainford, and well
wrought on the fpot.   Iron ore, in large quantities, is obtained
near Lindle, between Ulverftone and Dalton, in Low Furnefs.   Copper mines in the North have been worked, but
without much fuccefs.

* Williams's Natural Hiftory of the Mineral Kingdom, vol. i. p. 171.

CHAPTER II.

STATE of PROPERTY.

SECT. 1.—*Eſtates.*

SINCE the introduction of manufactures, property has become more minutely divided. But there remain proprietors who ſtill hold very extenſive poſſeſſions.

The remark made by Camden, in his BRITANNIA, of the number of ancient families bearing the names of the places where they reſide, and whence they took their names, is ſtill applicable to this county, *e. g.* Atherton, Bold, Fazakerley, Formby, Hoghton, Hulton, Mawdſley, Townly, Trafford, &c.

The yeomanry, formerly numerous and reſpectable, have greatly diminiſhed of late, but are not yet extinct; the great wealth which has in many inſtances been ſo rapidly acquired by ſome of their neighbours, and probably heretofore dependants, has offered ſufficient temptation to venture their property in trade, in order that they might keep pace with theſe fortunate adventurers.

Not only the yeomanry, but almoſt all the farmers, who have raiſed fortunes by agriculture, place their children in the manufacturing line.—The farmers in this county moſtly ſpring from the induſtrious claſs of labourers, who, having ſaved by great economy a ſum of money, enter upon ſmall farms, and afterwards, in proportion to the encreaſe of their capitals, take larger concerns. Nothing appears more deſirable to the proprietors of large eſtates, than that many cottages ſhould have annexed to them *a few* acres of land, which ſerve as a ſchool to the occupier in Agriculture, by giving his mind an opportunity of being employed in the management of it. An ob-

8   ſerving

ferving proprietor may always felect from amongft them proper tenants for his fmall farms, who may rife, as numbers have
done, to occupy with advantage the largeft farms in their
neighbourhood.

Eftates are principally under the direction of ftewards and
bailiffs. A few individuals have attended perfonally to the
improvement of their own lands; and having executed their
work in a fuperior manner, without doubt have found their
account in the fuperior profit derived from fuch exertions.

SECT. 2.—*Tenures.*

THE Tenures are chiefly freehold. There are fome
copyhold; leafes on lives have been more frequent formerly,
than at prefent; but the practice for granting leafes for lives is
not entirely difcontinued. A confiderable eftate in the county,
is poffeffed by the tenants in the following manner, which may
give fome idea of the proportion of leafes for lives, compared
to thofe for a determinate period.

	Statute Mea-fure.			Prefent Rent.			Eftimated clear Yearly Value.		
	A.	R.	P.	£.	s.	d.	£.	s.	d.
Amount of Leafes for one Life -	297	3	27	79	6	5	515	0	0
Ditto - - - for two Lives -	322	3	19	30	14	7½	458	13	0
Ditto - - - for three Lives -	222	0	13	80	2	4	352	3	0
Ditto - - - for years - -	1,742	2	4	3,910	2	8	4,212	1	2
	2,495	2	1	4,100	6	0½	5,537	17	2

CHAPTER III.

BUILDINGS.

SECT. I.—*Houses of Proprietors.*

THERE are not many noblemen's feats in the county, and thofe have been already fufficiently defcribed in former publications.—In regard to the feats of the gentry, Ince Hall, belonging to Henry Blundell, Efq; deferves particular notice, having been much improved by its prefent poffeffor, not only by the addition of excellent offices, hot walls, green-houfes, and paddock, but alfo on account of its being ornamented with many excellent paintings by ancient and modern mafters, foreign and Englifh; many marble ftatues, rich tapeftry, and other articles, have been felected to embellifh this feat, in a ftyle fo as to become a place of refort to the young artift, the virtuofi, and the curious traveller.

The buildings of the merchants and tradefmen difperfed over many parts of the county, and particularly in the vicinities of large towns, certainly merit notice *.—Confiderable expence having been laid out of late years in the erection, finifhing, and embellifhing many of them in a fuperior ftyle—many of which are furnifhed with hot walls, green-houfes, the rareft plants and fineft fruits: the adjoining grounds have been improved, laid out in various ftyles, and fringed with plantations.

* Some gentlemen tradefmen's houfes in Manchefter are erected upon a fcite of land, which pays a fum no lefs than 50 *l.* per annum ground-rent.

A tafte

A tafte for the fine arts has alfo gone forth, and a great number of expenfive paintings have been purchafed to ornament the walls, and engravings to fill their port-folios.—There are more readers amongft the lower clafs of people, it is fuppofed, than in any part of the kingdom.

S E C T. 2.—*Farm Houfes and Offices;*
and Repairs.

S o m e of the old built farm-houfes are ill conftructed; and (which may appear extraordinary, in a county where flate abounds, and ftraw fells at an advanced price) are ftill thatched, and the preparation of the ftraw for thatch is but ill managed. Fern is faid to make the beft covering, being naturally dry, and not apt to ferment like ftraw.

The more modern buildings, belonging to the Earl of Derby and many others, are ufeful conftructions; and in general fufficiently fpacious to contain the crops both of hay and corn.

Farms of fixty pounds a year, in Lancafhire, have offices frequently as large as would be thought to fuffice, in other counties, for farms of three or four hundred *per ann.* where it is the cuftom to ftack their corn, which is not the general practice in Lancafhire.

Mr. Boyer, of Lathom houfe, has favoured the furveyor with the following plan of a farm-houfe and offices, which have been lately erected upon Mr. Bootle's eftate.

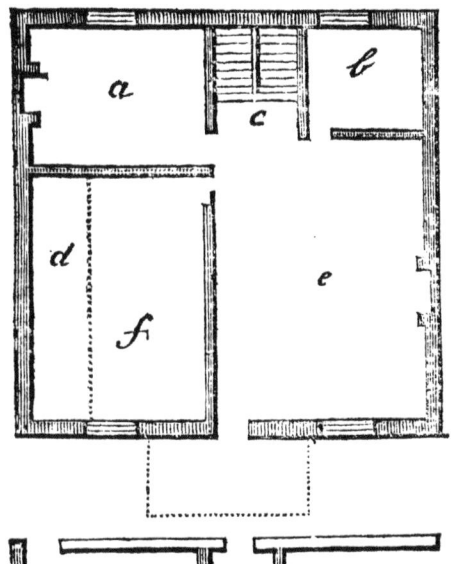

HOUSE.

a. Parlour.

b. Dining-room.

c. Staircase.

d. Milk-house.

e. Kitchen.

f. Pantry.

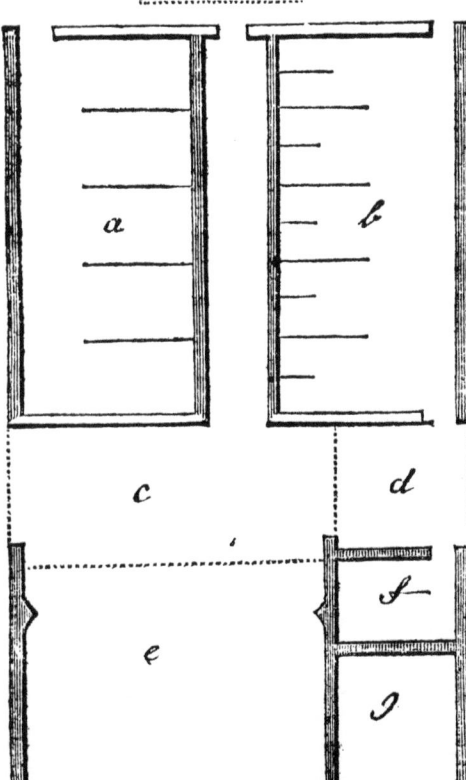

OFFICES.

a. Stable.

b. Shippon, or cow-house.

c. Thrashing-bay, or barn.

d. Shed.

e. Corn-bay, or barn.

f. Calf-crib.

g. Cart-house.

———

In forming buildings, convenience is the first thing to be considered : this is the only plan that has yet been erected in this county upon this construction; it will soon be allowed by the intelligent farmer, that two front parlours and front door are unnecessary, two out of the three are seldom or ever used. It is easily conceived from the plan annexed, how much more room is gained, and with what snugness every apartment joins. Many a farmer s wife would think herself in paradise in such a kitchen—these buildings are large enough for

D   farm·

Repairs have been frequently left to the tenant, as by covenant ftipulated; but they are in general fo ill performed, under
thefe terms, as to prove no fecurity to the landlord.   Some
landlords provide materials, and the tenant is to cart them to
the place wanted; which practice feems the only fervice in repairs by the tenant that proves of real utility to the landlord.

S E C T. 3.—*Cottages.*

THE cottages are of different kinds, for the common
labourer in agriculture, and the different artificers, and with
accommodations accordingly, with feparate rooms, to hold the
utenfils of trade in their various occupations ;—near large fac-
tories, being frequently built in long ranges adjoining together, and near the works, and fometimes accommodated with
fmall gardens *.

farms of £. 100 a year, but may be reduced or enlarged at pleafure. The
houfe ought to ftand with the kitchen and pantry to the north : a porch is
alfo neceffary, and ferves as back kitchen, to put milking veffels in, &c.
and break off the north-weft winds.
    The fet of offices may be varied many ways, and alfo be ufeful and conve-
nient; the one laid down appears to be the beft adapted for farms not exceeding £.100 a year ; it will hold five horfes and ten cows : you pafs to fuckle
the calves under cover, and they are clear and convenient to the fhippon, for
it is well to keep the fucklings at a diftance from the cows. Over the calf-crib
and cart-houfe is the granary, into which you afcend out of the barn, and
have no fteps to the outfide of the building; you may load the cart under
cover with eafe through a trap-door into the cart-houfe.   In large farms I
would add a ftable to one end, and convert the prefent ftable into another
fhippon, or cow-houfe.

    * Where the cottager has a fmall garden, the following mode of laying
potatoes may be of particular ufe to him :
    From every eye in each potatoe fet, will proceed different ftems ; which
when they are about nine inches above the furface of the ground,
fhould be fpread out in a circular form, bent down, and covered all over
(but juft the ends) with earth.   The following rude fketch may probably
render it more intelligible : a pit of earth nine inches diameter,
about one foot deep, dunged, then covered with a little mould,
upon which is depofited the potatoe whole, that is uncut. From
this fet may arife feveral ftems, which when of length fufficient, then the ftems bent down thus : and from the ftems
thus covered a few inches deep, and rounded up in the
fhape of a mole-hill, new fibres will ftrike, take root, and
potatoes be produced in large quantities.
    This mode may be ufeful to the cottager, as the prac-
tice requires but little dung, fome additional labour ; but
as the pits may be varied, the fame ground may be repeatedly and repeatedly planted.

Many

Many of thefe kind of dwellings have been erected by building fpeculators, as they generally (if the rents are paid) are
calculated to yield an intereſt of £. 10 *per cent.* to £. 20 *per cent.*
The modern buildings are chiefly of brick, and covered with
flate ; rents from one to five and fix pounds *per annum.* The
old cottages of the county, of which fome are yet remaining,
are of unhewn ſtone, or poſt and plaiſter, clay floors, and
thatched roofs.

The gardens attached to fuch cottages are found of the
greateſt utility, both for the means of healthy exercife they
furnifh, and as enabling the cottager to raife a confiderable
quantity of food at a fmall expence.

Chapter IV.
MODE OF OCCUPATION.

Sect. 1.—*Farms.*

IN moſt townſhips * there is one farm, ſtill diſtinguiſhed
by the name of the Old Hall, or Manor Houſe (the reſi-
dence formerly of the great proprietor of that diſtrict) which
is of larger extent than any of the adjoining or neighbouring
farms. Few of thefe farms, however, exceed 600 ſtatute
acres ; many do not extend to the amount of 200. But the
more general fize of farms is from 50, 40, 30, down to 20
acres a-piece; or even fo much only as will keep a horfe or
cow only; or one of thefe, as is moſt convenient.

Farmers in general are charged with being ſtupid, obſtinate,
and attached to old cuſtoms. In this county they do not
altogether merit thefe harfh accufations — we have all our
prejudices and attachments. They are, in general, a laborious, and certainly a moſt ufeful clafs in fociety. The
hazards they have to encounter, from feafons, and other
caufes, leave little room for trials of uncertain experiments.
After the grain has been depofited in the earth, the ground
being previoufly prepared to receive it, in the moſt hufband-

* The parifhes of Lancaſhire are again fubdivided into townſhips.

man-

man-like manner, ſtill the ſucceſsful iſſue entirely depends on a favourable ſeaſon to vegetate and mature the grain. Mildews and blights, under theſe favourable aſpects, may yet intervene; but ſhould not any inauſpicious appearance happen, and ſhould the reaper be prepared to gather the produce of the loaded fields, yet how often does the howling blaſt ſcatter and diſperſe the hopes of the huſbandman!

Again, the labours of the farmer are toilſome; his gains cannot be great, upon the moſt favourable calculations ; namely, that from his grounds he ſhould be enabled to raiſe three rents—one, of courſe, his landlord demands ; more than another is requiſite to maintain his family, pay the hire of ſervants, and ſupport contingencies; the third and laſt, toward paying intereſt of the capital advanced for ſtock, and afford an annual ſurplus to reward his labours.

A ſpirit of ingenuity and improvement amongſt the inhabitants of this county, has been frequently proved, and is yet, every day, manifeſting itſelf; but this is moſt apparent amongſt the manufacturing claſs; and the reaſon is obvious —reward immediately enſues. The Glaſgow manufacturers, till of late, have exceeded the Lancaſhire in muſlins. Stimulated by emulation, in the neighbourhood of Bolton, they now boaſt that they have at laſt, and but very lately, ſurpaſſed the Glaſgow muſlins and fancy-works. The ſame flame would equally ſhine amongſt the farmers as well as amongſt the manufacturers, were the reward equally certain: ſtill it remains to enquire how a ſpark of this flame may be kindled. The farmer is not ſuch a novice, and ſo totally blind to his own intereſt, as to be incapable of viewing the effects of ſkilful cultivation, however novel; and if on repetition this new practice be found beneficial, the great incentive to action, INTEREST, will operate equally upon one individual as on another.

But how is the farmer to be convinced? He is told ſuch are the cuſtoms which ſucceed in other diſtricts; but theſe aſſertions do not convince. Soil, climate, or other cauſes, may operate — he waits an example nearer home. Herein the landlords, the gentlemen of property in the county, ſhould interfere and ſet the example; and ſeveral ſpirited gentlemen

have

have made great exertions in the introduction of many novel practices, and under great disadvantages; for not being able to execute, but only to direct, they have had both prejudice and ignorance to encounter. Their labour is always procured upon worse terms, probably by £. 50 *per cent.* than can be obtained by the farmer or gardener, who can say to his workmen, " *Come, let us go dig together*;" even if the labourer be hired to work by the piece upon the usual terms, it will often be slightly performed.

How many good effects, and what superior cultivation, has been produced within the space of half a century; by these means, in a slow and almost imperceptible manner! The very village in which this account is written *, half a century ago, was not able to supply from its own meadows an inferior number of cattle, with a sufficiency of hay for winter stock. What was wanted of this article was generally purchased from the Sefton meadows.

There is a greater quantity of live stock at present kept, and yet no small surplus of hay remains to be sent to the Liverpool market.

It was in the memory of a worthy and experienced farmer †, who only died the present year, that the first load of night-soil brought from Liverpool towards the north was by his father; who was paid for carting the same the price that heretofore had been paid for carting away this nuisance, and throwing it into the river Mersey.

The good effects upon the land, which experience has proved dung to have, have caused it, at this period, to be sold at an advanced price, and carted to a considerable distance. The varieties of potatoes, their diminished value to the purchaser, in comparison to the price they fetched twenty or thirty years ago, under an advance of land, dung, and labour, proves superior cultivation, and much greater produce of this excellent vegetable, from the same quantity of soil. The introduction of clover, the varieties of seeds of grain, both oats and wheat, prove some degree of atten-

* Walton, near Liverpool.
† Mr. John Harper, late of Bank Hall.

tion;

tion; as does alfo the introduction of the turnip, although
the cultivation at prefent be not fo extended, nor treated in
the moft hufbandman-like manner.   Yet this, and all the
above examples, are introduced to prove that a Lancafhire
farmer, though not a complete agriculturift, is not without
fome fpirit of improvement.

S E C T. 2.—*Rent.*

THE rent of lands is very variable in the different parts
of the county, from ten fhillings to ten pounds *per annum,*
the large acre, of eight yards to the rod; the latter enormous
fum, being frequently paid in the vicinity of large towns, for
particular accommodation. The price paid by the farmer is from
ten fhillings for fome barren lands, up to twenty, thirty, forty,
and fome (but not many) as high as eighty fhillings *per* acre
per annum (large meafure.)

S E C T. 3.—*Tythes.*

THE tythes are in many places collected, one *eleventh* of
the corn—the hay is frequently converted, five fhillings *per*
acre for old meadows, fix fhillings *per* acre for firft year's clover (acre large meafure.)

S E C T. 4—*Poor Rates.*

POOR rates are at Liverpool 2 *s.* 6 *d.* *per* lb.; at Walton
12 *d.* *per* lb.; at Manchefter 6 *s.* *per* lb. at a highly-valued
rental, but taxed at only half value, they are therefore at 3 *s.*
per lb. on the rental; at Bolton 6 *s.* a late affeffment, but
would be 4 *s.* of full and prefent value; Rochdale about 4 *s.*;
at Weft Houghton 16 *s.* the pound; Afhton 5 *s.* *per annum;*
Oldham 5 *s.* *per annum* on the full rental.

S E C T. 5.—*Leafes.*

MANY farms are held by leafes on three lives *, on which
a fine has been paid, and a fmall annual rent referved; and

---

* When a leafe is granted for three frefh lives, on an average the term
lafts upwards of 50 years.

fometimes

ſometimes an addition of *boon ſervices*; which laſt ſyſtem ſeems much on the decline. Theſe leaſes are generally eſtimated at about fourteen years purchaſe.

The leaſes upon years are, from ſeven, eleven, to fourteen; but chiefly ſeven *. Covenants in ſome to pay the rent the day the tenant enters upon the premiſes. This covenant for the ſecurity of the landlord, but not exacted except on emergencies. The time of entering upon the lands is Candlemas; and on the buildings, May-day. Uſual covenants are, the landlord to repair buildings, the tenant carting the materials. The tenants ſeverally to diſcharge all taxes, ſerve all offices, and all the duties charged upon the farm.

Tenants are reſtrained, by covenant †, to the quantity allowed to plow, ſometimes to one-third, ſometimes to one-fourth, of the whole; and alſo, of late, to the number of crops to be taken at one breaking up of the ground—ſometimes to our crops; and ſometimes only three are allowed. Tenants are reſtrained, by covenant, from ſowing wheat upon bean ſtubble ‡, or any other ſtubble from which a crop has been

taken

---

* Short leaſes, where farms are arable, and upon an improving plan, ſuch as marling, and any other ſort of improvement, manure, &c. being ſo dear, would greatly check the ſpirit of improvement, whence of courſe in time the land muſt decreaſe in its value, when farmers have to do at their own expence. If the proprietor was to improve at his own expence, and finds there is an advantage in ſhort leaſes, it is but juſt ſo to do; but when the farmer has to improve at his own expence, and upon a ſhort leaſe, and cannot pay his way, the farmer is often ruined, and the proprietor a loſer; for when the farmer is poor, the farm is ſure to be made poor alſo.

Leaſes of a reaſonable fair length of time, and the covenants not ſo ſtrict at the beginning of the leaſe, would greatly encourage the ſpirit of the farmer to improve, when he has to do it at his own expence, and the covenants to he ſuch as to bind the farmer by forfeitures, over and above the yearly rent, ſo as to have the premiſes in a high ſtate of cultivation at the expiration of the leaſe, and to give his management of manures, &c. in writing, and his account of ſtock every year of the term to the proprietor or his agent, or as often as it may be required; and if found in an error, the covenants to be ſuch as to bind him in a ſum, according to the rent and largeneſs of his farm, over and above all forfeitures, and likewiſe to forfeit his leaſe.—*Mr. Harper.*

† Tenants are *very much reſtrained* in plowing, and not improperly, unleſs they could be induced to cultivate green crops, as turnip, cabbage, &c.

‡ Bean ſtubble, in ſome counties, is almoſt the only and beſt tilth known for wheat; and probably, with proper culture, might be ſo in this
county.

taken the fame year. The tenants, by covenant, reftrained from paring or burning, except mofs lands.

The tenants fometimes reftrained, by covenant, from felling either hay or ftraw, but are bound to confume the whole upon the premifes.

The tenants, by covenant, reftrained from felling off their ftock till the clofe of the year, at the expiration of their term, that the greater quantity of dung may be raifed from the produce confumed.

The tenants allowed to take off three-fourths of the wheat growing upon the premifes at the expiration of a leafe. The fucceeding tenant to have the remaining quarter *.

A fucceeding tenant to have permiffion, after Candlemas, at the expiration of a leafe, to occupy certain portions of the out-buildings, by claufes founded for the accommodation of his horfes, hay, &c. neceffary for the fpring feeding, on the new tenant entering upon his farm.

Upon the eftate of that intelligent landlord, Mr. Bayley, of Hope, whenever a tenant wifhes for the whole of his farm, or any particular field, to be improved, by draining, marling, liming, dunging, or laying down to grafs in a fuperior manner, the landlord takes the field into his own poffeffion, during the procefs; and, when completed, returns it again to the tenant, with an advanced rent of ten *per cent.* upon the capital laid out upon the improvements; by which fteps Mr. Bayley has advanced the rental of his eftate, fince the year 1768, very confiderably—his

county. At any rate it is wrong to prevent the tenants from making trials, and ftill more fo to prevent them making one of the firft of agricultural improvements on poor land, and turf-burning.—*Mr. Boys, of Kent.*

This covenant, to the ear of a Kentifh farmer, feems the moft extraordinary that can be. It is by them fuppofed to be almoft the beft preparation that can be for wheat. Nothing can juftify it, but inattention to the bean crop, by which the land is in too foul a ftate to be fown : but it muft be foul indeed, if a Kentifh farmer could not find ways to clear it properly for wheat. No man can be furprized that corn is fo dear in this county, when he knows that fuch covenants are to be found in leafes.— *Mr. Dann, of Kent.*

* The cuftom of the off-going tenant taking three-fourths of the wheat left growing on the premifes is much upon the decline in this diftrict, for it is generally allowed but one half; and many leafe not to leave any at all. I myfelf, for one, am not to leave any growing.—*Mr. Harper.*

tenants are thriving, and getting money.   Mr. James Balmer, who accompanied the furveyor in this excurfion, and is a good judge of cattle, declared he never faw, upon any one eftate, fo large a ftock of cattle, uniformly good, being the Lancafhire long horn, and what he termed the right fort.

A certain method to excite improvement would be to let farms to men of induftry, ingenuity, and property, upon reafonable terms, and give leafes for 21 years, free from arbitrary covenants; without this nothing can excite a general and effectual improvement.   For suppofe a farmer to lay out a few fcore or hundred pounds upon his farm in ufeful improvements; his landlord fees the advantage he is making, fends a valuer to look over his farm; who, never confidering (nor being told) what he has done, lays a tax upon his induftry, and makes him pay intereft for his own money.   Daily experience proves the truth of this affertion, and will ever operate to the deftruction of improvement, and of courfe to the great difadvantage of the public.

Another improvement which here fuggefts itfelf, is by a revifal of the covenants in leafes; and adapting them better to the prefent improved fyftem of agriculture; many of them at prefent militate againft fome approved practices, nor has an ingenious cultivator fcope to act, being reftrained under covenant.   There wants a fpirit of liberality in the general tenor of leafes.   Inftances might be produced to prove that if indulgencies were upon fome occafions granted, the tenant would be benefited, and the landlord enriched; and this only by a new modelling of the covenants, whereby the lands, if managed under a certain cultivation, muft return, at the expiration of the leafe, into the hands of the poffeffor, in a better ftate than they were in at the beginning; and of courfe, would bring an advanced rental to the eftate; and again, the tenant, if induftrious, might be enabled, by his advanced capital already gained by his former leafe, and the fuperior ftate in which he now finds the lands upon the fame farm, to give the advanced rent to his landlord with greater profit to himfelf, than upon his former rent, under the impoverifhed ftate in which farms are generally entered upon.

E

Leafes upon lives only act as checks to improvements; they are, in general, only beneficial to the firft purchafer, who fecures an income on three lives, for fourteen years purchafe—the fee fimple of which would have required double the fum. The fucceffors, elevated by poffeffing an eftate under a fmall annual quit rent, inftead of full rent, *live up to the height*, as the phrafe is, and are but ill-prepared to renew the leafe, or pay the fine required when a life drops. The leafe, through inability of the tenant to renew, or fome other caufe, is fuffered to run out, under the uncertainty of life, and the lands (there being no provifion made by covenants to prevent it) are haraffed and abufed to fuch a degree, as to require a length of time to reftore them.

Theory and practice, it muft be confeffed, are perpetually at variance, as well in Agriculture as many other purfuits. It might at firft fight appear, that the cuftom of granting leafes for three lives (a tenure that gives fuch probable fecurity to a tenant) would excite a degree of fpirit of improvement amongft the holders of thefe tenures. Experience however proves the contrary fact—For leafeholds upon lives are generally under the moft wretched cultivation.

Eafy rents may have produced a carelefs indolence, and hence an averfion to enterprize. The landlord having but little intereft in fuch eftates, and lefs power over fuch tenants, is himfelf checked from any fpirit of improvement upon fuch contingent property. Thofe proprietors who look a little towards the welfare of pofterity, are come to a refolution of running thefe tenures out, and, of courfe, the tenants are not behind in exhaufting and every way impoverifhing the land.

The ancient cuftom of granting leafes for three lives is beginning to difappear: It fhould feem probable that this tenure, which grants fo much fecurity to the tenant, would naturally excite a liberal and enterprifing fpirit of hufbandry: fact however proves the reverfe of the propofition; the ancient leafehold eftates being almoft univerfally in a wretched ftate of cultivation, beyond all comparifon lefs productive than thofe held upon fhorter tenures. Eafy rents, fecure poffeffion, and good

land,

land, have lulled the leafeholders into a carelefs indolence, an averfion to enterprize, which have been productive of much ill to themfelves and their connections, and, above all, to the public; much ill has accrued to the leafeholders from the power of borrowing money upon this ideal fpecies of property.—Thefe obfervations hold good to the cuftom of half *rent* and half *fine*. Upon fuch tenures the immediate landlord can have no inducement to advance money for the amelioration of his eftate, and but little intereft and lefs power either to prevent his land being exhaufted by wretched hufbandry, or to oblige his tenants to keep upon their farms a due proportion of ftock. Whoever will take the trouble of examining the eftates of this county held upon three lives, will find the arable worn out by a perpetual fucceffion of exhaufting crops, and the grafs little more than a collection of rufhes and beggary, the whole un-ditched, undrained, and unmanured. Landlords have at length become fenfible to their own interefts, and are fuffering their leafes to run out, which, though a wife policy, is deftructive in its immediate effects: in fact, the country is at this time fuffering extremely in confequence.

Modern leafes upon land in high condition are from feven to eleven years;—upon improveable land fourteen to twenty-one: —But landlords in this county will never adopt the fyftem of granting *long leafes* free *from all reftrictions*, fuch as are recommended by the furyeyors for the Weft Riding of York.— To recommend fuch a fyftem to a manufacturing county would be abfurd.

The firft purchafer of leafeholds is generally a fenfible induftrious man, who underftands his bufinefs, and attends to it. His fucceffors are often both ignorant and idle, but their tenure is fecure, and they cannot be difturbed in their poffeffions by any thing but their own folly; this often induces them either to harafs their eftate themfelves, or let them off at rack rent to fome poor devil, without any capital or means of procuring one.

E 2

I know

I know that the contrary may be, and often is the cafe, and that the abufe of a good cuftom is no argument againft the cuftom itfelf: but I alfo know that there are no poorer or more wretched people in the county than the occupants of leafehold eftates, and that the fons and grandfons of moft of the original leafeholders are not to be found upon fuch eftates —A middle man is the devil—all the-world knows the confe-quence of this cuftom in Ireland—*the little lords* of this coun-try are in the fame predicament.

Differences between Landlords and Tenants.

The juftices might fettle all differences * and difputes be-twixt the landlords and tenants, inftead of the prefent expen-five mode of courts of judicature. The differences are gene-rally of a trifling nature, and eafy to be comprehended. The tenant would be more likely to obtain redrefs under this mode of judicial enquiry, and the landlord would prevent abufes to his land : he may now be withheld, under certain circumftances, from correcting a refractory tenant, which might be too heavy for any redrefs the landlord could obtain ; and the damages given too grievous for a tenant to bear.

S e c t. 6.—*Expences.*

Authentic Statement of a Farm, communicated by Mr. Henry Harper, of Bank Hall.

	£.	s.	d.	£.	s.	d.
Yearly Rent - -	270	0	0			
Taxes - - -	45	0	0			
				315	0	0

---

* With a *proper jury* perhaps they might. But how would fuch fum-mary proceedings operate on the pockets of a moft numerous tribe in this country, the gentlemen of the law ?—*W. D.*

Outgoings for one year	£.	s.	d.	£.	s.	d.
Sugar - - -	7	10	7			
Tea - - -	1	17	5			
(1) Bread - -	4	7	2			
(2) Butchers meat - -	6	0	11			
Currants - -	0	19	9			
Vinegar - -	0	8	4			
(3) Soap - - -	1	19	8			
(3) Candles - - -	1	5	11			
Starch - -	0	2	5			
Mugs - -	0	7	6			
Flax and wool - -	3	2	0			
Salt - - -	2	12	8			
Malt and hops - -	6	10	0			
Liquors - -	5	10	0			
Cheese - -	31	10	0			
Coals - -	6	16	0			
				81	0	4
Nine servants' wages,						
who live in the house -	66	0	0			
Three labourers - -	78	0	0			
Mowing and haymaking -	20	0	0			
Reaping corn - -	12	0	0			
Manure - -	100	0	0			
* Tythe hay - -	7	10	0			
Blacksmith - -	25	0	0			
Wheelwright - -	25	0	0			
Collar-maker - -	5	0	0			
Cooper - -	1	0	0			
				339	10	0

(1) Wheaten bread purchased for tea, and on extraordinary occasions.
* Corn is collected (tythe) in kind.

	£.	s.	d.			
				£.	s.	d.
(1) Provender - -	50	0	0			
Farrier - - -	3	3	0			
Miller - - -	3	0	0			
Weeding - - -	2	0	0			
Repairs of gates and building	5	5	0			
Timber for rails and fences	3	3	0			
Mole-catcher - -	2	12	6			
Rat-catcher - -	0	10	0			
Store pigs - - -	3	3	0			
(2) Cloaths and extra expences	3	10	0			
Lofs in cattle - -	20	0	0			
Bad debts - -	20	0	0			
				144	6	6

£.879 16 10

Stock kept upon the Bank Hall from March 1794, valued by a fworn appraifer, for the ufe of the Lancafhire Report.

	£.	s.	d.
Ten draught horfes, 15 l. each -	150	0	0
Three three-year old colts, 15 l. each	45	0	0
One two-year old colt - -	10	0	0
Two year old colts - -	10	0	0
A hack horfe - - -	10	0	0
A poney - - -	3	3	0
Twenty-five milch cows, at 7 l. 10 s. each - -	187	10	0
Seven in calf heifers, at 8 l. each	56	0	0
Nine heifers barren, 4 l. 10 s. each	40	10	0
Fourteen one year old, 3 l. 10 s. each	49	0	0
Ten rearing calves - -	10	0	0
Two bulls - - -	15	0	0
One brood mare in foal - -	20	0	0
	606	3	0

(1) Malt-duft, bran, &c. purchafed to give to the cattle, mixed with potatoes or turnips.

(2) Mr. Harper is a bachelor ; his family confifts only of himfelf and nine fervants.

Three

	£.	s.	d.		£.	s.	d.
Three carts, at 15 *l.* each - -	45	0	0				
Three smaller carts, at 6 *l.* each -	18	0	0				
One water-cart, pump, &c. for con-							
veying water from dunghills -	15	15	0				
Three single ploughs, 15 *s.* each -	2	5	0				
Three pair harrows, 15 *s.* each -	2	5	0				
One large harrow - - -	1	10	0				
				84. 15	0		
Four sets of horse gears for 3 horses	16	0	0				
Nine spades, at 2 *s.* each - -	0	18	0				
Twelve dung forks, at 1 *s.* 3*d.* each	0	15	0				
Twenty-four pitch forks, 6 *d.* -	0	12	0				
Twenty rakes, at 3 *d.* - -	0	5	0				
Forty chains and hoops to fasten cattle	1	0	0				
Marling hacks and hedging tools	0	12	0				
Marling piles and lumber - -	1	0	0				
Two wheelbarrows - -	0	18	0				
Dairy utensils - -	8	0	0				
Winnowing machine - -	3	3	0				
Forty sacks, at 1 *s.* 3*d.* - -	2	10	0				
Riddles and sieves - -	0	5	0				
One bushel measure - -	0	7	6				
One half bushel and peck - -	0	7	6				
One winnowing sheet - -	0	5	0				
One machine for cutting straw -	1	5	0				
Six cart ropes for binding - -	0	18	0				
Harrowing geers - - -	1	0	0				
Thrashing machine - -	50	0	0				
Building for gangway - -	30	0	0				
				120	11	0	
				.811	9	0	

N. B.—The houshold goods not brought into this account.

	£.	s.	d.
Rent, with taxes - -	315	0	0
Outgoings - - -	564	16	10
Intereſt of ſtock - -	40	11	5 ¼
	920	8	3 ¼

	£.	s.	d.
Three rents, with taxes - -	945	0	0
Three rents paid to the landlord -	810	0	0

The Bank Hall eſtate was, in the year 1793, under the following cultivation.

The acres are given in the cuſtomary meaſure of eight yards to the rod.

Bank Hall eſtate - - -	69	acres
Bootle Marſh, improved 1780 - -	53	
Bootle Marſh not yet improved -	25	
	147	

Diſtribution of crops.

Old meadow for hay - - -	40	acres
New meadow, clover and graſs ſeeds	12	
Wheat - - -	10	
Barley - - -	6	
Oats - - - -	6	
Beans - - -	3	
Potatoes - - -	2	
Fallow - - - - -	8	
Turneps - - - -	2	
Paſture - - -	33	
Paſture, unimproved land - -	25	
	147	acres

Chapter

CHAPTER V.

IMPLEMENTS.

USEFUL INSTRUMENTS IN HUSBANDRY.

ABOUT thirty years ago, the Rotheram or Yorkſhire plough was introduced into the ſouthern part of this county.*. The plough formerly in uſe was almoſt a load of itſelf for a draught horſe. In the north of Lancaſhire a plough, called the Cumberland plough, invented in that county, is generally uſed. A trench plough has been lately introduced by Mr. Ducket, ſon of the celebrated Ducket of Eſher, in Surrey.

This plough has a ſkim coulter, by which the ſurface (if foul) may be turned under, and freſh ſoil brought up; as it is capable of bringing up the land from ſix to ten inches, and is uſually drawn by three horſes. Another inſtrument has been lately introduced, which Mr. Eccleſton, with propriety, calls the *miner*; which is a plough-ſhare fixed in a ſtrong beam, without mold-boards, and drawn by four or more horſes, and follows in the furrow the plough has juſt made, and, without turning up the ſubſtratum, penetrates into, and looſens the ſoil, from 8 to 12 inches deeper than the plough had before gone; which operation, beſides draining the land, cauſes the water to carry along with it any vitriolic or other noxious mat-

* By the late J. Atherton, Eſq. Walton Hall.

F   ter;

ter; by the fubftratum thus loofened, the roots of plants may penetrate the deeper; and, in courfe of time, that which is but a barren fubftance may become fertile foil.

There is a greater variety of carts in this county than in the fame given fpace in any other part of the kingdom. In the neighbourhood of Liverpool they are of very large fize; thofe employed in the coal-trade within the town are gauged to 36 bufhels Winchefter.

The country dung carts, in the fame neighbourhood, are alfo of a very large fize, and generally will hold thirty-fix Winchefter bufhels, and carry two tons of dung; they have fix-inch wheels. In the interior parts of the county, the carts greatly diminifh in fize, and have variety of forms; in the northern part the fize is very fmall; the clog wheel, as it is termed (three planks of afh); which was formerly much in ufe in the north, on account of cheapnefs, has yielded to the fpoke wheel; the clog being more clumfy, and the cart more liable to overfet—in thefe carts the wheel did not move upon the axis, but both turned round together.

Single carts are in more general ufe*. Mr. Jenkinfon of Yealand fays, " that a gentleman, in his neighbourhood, made a fair trial in the hay field between the large and fmall carts, or what is often called double and fingle carts, in which the latter had much the advantage, in difpatch of bufinefs; and the confequence was, that the double carts were little ufed afterwards." Mr. H. Harper obferves, that for a fmall diftance, e. g. half a ftatute mile, the fingle cart has certainly an advantage; but at a further diftance, he prefers the double-cart for difpatch of bufinefs; becaufe the fame ftrength or number of hands is requifite to unload the fmall, as the large cart.

Although Lancafhire is not a corn county, yet, labour being dear, there are feveral thrafhing machines already introduced; one of which belongs to Colonel Mordaunt of Halfall, which

* The encouragement of thefe can alone preferve the roads of Great Britain.—See Annals of Agriculture, Vol. XVIII. p. 178, &c.

moves

moves by water, thrashes, winnows, and grinds (or crushes the corn for provender), all at the same time. Many of the neighbours apply to this machine, for the use of which the colonel takes or charges one twentieth part *. Hand machines are also introduced, and are useful to the farmers, chiefly made by John Naylor, at Ashton, near St. Helen's †.

This machine requires two men to turn, a boy or girl to feed, and another to take away the straw ‡. The price of these hand machines are about six pounds each.

A churn has.been lately introduced, which seems very useful for its neatness, cleanliness, and economy (as it occasions less waste of milk). The churn, or vessel, instead of being round, has four corners, and the milk is put in motion by turning a handle, upon which are fixed boards which move horizontally in the manner of a reel within side the vessel,

---

* The average price paid for thrashing in the district.

† On this it is remarked, on the margin of one of the Reports, that " The surveyor has been wrong informed of the above; for I have seen three or four different ones of the said John Naylor's make, and never any that would do as much as the information given. From all the enquiries I could make of them, there was little or no benefit arising therefrom beyond thrashing with flails. One of his make, in particular, I saw at work at the honourable Mr. Jones's, of 'Blackley-hurst, in Billinge, from which I could see no advantage.

" Mr. Steevens of Chorley, who erected Colonel Mordaunt's of Halsall, made one for Mr. Blundell of Ince, to be worked by hand, which would have taken as many men to have worked it as would have thrashed the corn with flails; for which reason it was returned to the maker as per agreement. Mr. Steevens I have since seen, who said it was impossible for one to be erected to any advantage to be worked by hand; it must either be by water, steam, or horses."

‡ Hire of two men   -   -   -   0   3   4 who can, with this machine,
  Boy and girl   -   -   -   0   2   0 thrash about 30 bushels
                                 ───────── of wheat *per* day, which
                                 0   5   4

(*£. s. d.*)

would come to 11 *s.* 3 *d.* at the present price paid for thrashing: about 70 bushels of oats also *per* day, which would cost 9 *s.* 6 *d.* according to the present price paid. But hand machines will be found insufficient for this heavy work. Mr. Henry Harper, of Bank-hall, has contracted for a thrashing machine, to be made for 40 *l.* which will require one horse, and is to thrash out from eight to ten bushels *per* hour, according to the length of straw, and quantity of grain contained.

by

by which the operation of churning is fomething eafier, and the work expedited.

A hay-cutter, in the form of a fpade, ftraight, and fharp at the point and upon both fides, performs the work with much more eafe and expedition than the common hay-knife. This tool was introduced from Yorkfhire by Mr. Ecclefton.

The following is a Defcription of MR. HARPER'S *Mill; drawn up by himfelf.*

" MY thrafhing mill will thrafh all kinds of corn, and any kind of fmall feeds, perfectly clean*; and in wheat that is bad to thrafh by the flail, will get from two to four quarts out of a thrave, more than it is poffible to get out by the flail; and if it is good to thrafh, generally about a quart; and gets a deal more chaff of all kinds than the flail, which is ufeful upon a farm if properly applied, the merits of which belong to the mill; and it does not damage the ftraw more than the flail; and if the corn is not well got, it does not damage the ftraw fo much.

The following is the defcription of the mill : Firft is the horfe-wheel, which is fifteen foot diameter; in which there are 204 caft iron coggs, and at the end of a tumbling fhaft is a caft iron wheel, which contains 20 coggs, which work into the horfe-wheel; and at the other end of the tumbling fhaft within the barn is a fpor-wheel, feven foot and a half diameter, which contains 100 wood coggs, which work into a caft iron wheel or 18 inches diameter, containing 14 coggs, which is fixed on the end of the fhaft that comes through the cylinder. The cylinder is fix foot long and three foot diameter, and has fixed upon it twelve wrought iron beaters, all at an equal diftance; thefe beaters by the cylinder running round meet two

---

* The furveyor purchafed twenty thrave of barley ftraw, 1794, from a tythe barn, thrafhed with the flail in the ufual manner of the county, and at the ufual price paid. When he got it home, he caufed it to be thrafhed over again, and from this fmall quantity, which was made up into fmall fheaves, he obtained one and a half bufhel of good corn.

wood

wood fluted rollers five inches diameter and fix foot long, which are fixed fo as to be right in the center of the cylinder as it runs round; and under the cylinder is fixed a playboard, which is the whole length of the cylinder, and goes about one third of the way round the cylinder, and is made fo as to fit to the fhape of the round of the cylinder; and at one fide it is fixed clofe up to the under roller, and the other fide of the playboard is where the corn and the ftraw come out, only fome little that drops through the playboard, which is made with ribs of an inch and a half broad, and two deep, and half an inch diftant one from another: the playboard is faftened to the frame that fupports the machinery, where the corn and the ftraw come out on a fwivel, turned by two wood-pins; and at the other fide up to the under roller it is fupported by two wood levers, one at each end; which, by weight being hung on the levers, forces the playboard either nearer or farther off the cylinder: if the corn comes out foul-thrafhed, the play-board muft be forced nearer the cylinder by more weight.

The rollers are worked by the fpor-wheel, which has a caft iron wheel of two foot diameter, which contains 72 coggs, and is fixed right in the center of the fpor-wheel, which works into another wheel of the fame diameter and the fame number of coggs, which is fixed to a ftrong upright piece of wood that fupports the fpor-wheel, and this works into another caft iron wheel of one foot and a half diameter, and contains 52 coggs, which is fixed to the end of the under roller: thefe rollers draw the corn in from off a feeding board, which is as broad as the rollers are long; the top roller is fixed in wood levers, at each end one, which are faftened into the frame by a fwivel turn at one end, and the other ends of the lever lie loofe upon the frame, which have weights hung on them as occafion requires for drawing in the corn more or lefs.

Now as thefe different motions are all connected together, they appear to be fimple, and only take up eight foot fquare in the infide of the barn, by ten foot high. The gangway is eight yards fquare, which the horfe-wheel works in at the outfide of the barn. The horfes travel at the rate of 2 miles

and

and $\frac{7}{8}$ in one hour, and at that rate the beaters on the cylinder ftrike the corn 3,000 times in one minute: by the
motion of the machinery, the rollers draw the corn in at the
fpeed that the beaters ftrike it, full four times in one inch of
length of the corn drawn.

Expence of one day's work of eight hours of thraſhing corn,
&c. by the mill when kept in full woık.

To thraſhing 16 quarters of wheat.

To 3 horſes one day - at 2 s. 6 d.	0	7	6 £. s. d.
To 2 men ditto - - at 2 s. -	0	4	0
To 4 boys ditto - - at 1 s. -	0	4	0
To extra charge for winnowing -	0	2	0
			0 17 6

To thraſhing 24 quarters of barley.

To 2 horſes one day - at 2 s. 6 d.	0	5	0
To 2 men ditto - at 2 s. -	0	4	0
To extra man for one day - -	0	2	0
To 4 boys ditto - - at 1 s. -	0	4	0
To extra charge for winnowing -	0	2	0
			0 17 0

To thraſhing 32 quarters of oats.

To 2 horſes one day - at 2 s. 6 d.	0	5	0
To 2 men ditto - - at 2 s. -	0	4	0
To 4 boys ditto - - at 1 s. -	0	4	0
To extra charge for winnowing * -	0	2	0
			0 15 0

* The charge for winnowing ſeems trifling; but it ſhould be under-
ſtood, that when corn is thraſhed by the flail, and paid for by the ſcore,
the thraſhers always aſſiſt, without any additional expence paid for their
labour.

To

To thrashing 24 quarters of beans.

		£.	s.	d.	£.	s.	d.
To 2 horses one day	- at 2 s. 6 d.	0	5	0			
To 2 men ditto	- - at 2 s. -	0	4	0			
To 4 boys ditto	- - at 1 s. -	0	4	0			
To extra charge for winnowing	-	0	2	0			
					0	15	0
					3	4	6

Expence of thrashing different kinds of corn
on my farm by the flail.

		£.	s.	d.
To thrashing 16 quarters of wheat at 2 s. 10 d.		2	5	4
To ditto - 24 ditto of barley - at 1 s. 6 d.		1	16	0
To ditto - 32 ditto of oats - at 1 s. - -		1	12	0
To ditto - 24 ditto of beans - at 1 s. 6 d. -		1	16	0
		7	9	4
To expence of thrashing the same quantity of corn by the mill - - - -		3	4	6
		4	4	10

Hence it appears that the mill clears 120 *per cent.*

	£.	s.	d.
To expence of the mill, and fixing up - -	50	0	0
To ditto of gangway that contains the horse-wheel, which is at the outside of the barn -	30	0	0
	80	0	0
To the average of corn thrashed on my farm by the flail *per* year - - - -	30	0	0
To 120 *per cent.* saved by the mill, of thirty pounds - - - - - -	18	0	0
To 10 *per cent.* for 80 pounds, for interest and repairs - - - - - -	8	0	0
Clear *per* year towards paying off the principal -	10	0	0

The

The extra charge of two fhillings per day for every day's work of different kinds of corn, &c. thrafhed by the mill, for winnowing, is a charge that belongs to the mill.

For the thrafher that thrafhes the corn with the flail, either by the fcore or quarter, is always an affiftant hand to the winnowing, without any additional charge on the corn thrafhed.

Improvements on the thrafhing mill to be made. The firft of thefe I am under with my mill, for I have a man now making models to my own direction for caft iron wheels, with four different rows of coggs in every wheel, for to either quicken or flower the motion of the rollers that draw the corn in from the man that feeds it off the feeding board; for although the mill thrafhes all kinds of corn, &c. perfectly clean, let the corn be good or bad to thrafh, as the motion of the rollers are now fixed, it all gets thrafhed alike.

2dly. Gaining power in the horfe wheel, which will quicken the motion of the fpor wheel, and eafe friction from the horfes.

3dly. Small friction wheels for the cylinder to work off.

4thly. To have canvas fixed upon rollers, to draw the corn in more regularly.

5thly. Stones to be fixed to grind corn.

6thly. To cut ftraw.—7thly, to wafh clothes.—8thly, to churn, and pump water.

Now the ten pounds per year, that appears to be faved by the mill towards paying off the principal, as a farmer I do not mean it for that purpofe, nor to deprive the labourer fo much of his employ, but am happy in finding myfelf fo fituated as to get my corn ten *per cent.* cleaner thrafhed, and with fo much difpatch, and in fo little time that I can take my labourers to any bufinefs the farm may require, fuch as pruning fences, clofe or open draining; and fo much cafh that is faved by thrafhing, laid out yearly in the employ of labourers for that ufe on the farm, will pay me fifty *per cent.* better, and improve the farm more than keeping one man ten months in the year *batting* * in the barn, or even to half the time, and thrafhing with the flail. There is not one labourer in twenty but what would rather do any labour on the farm, than thrafh; and if he thrafhes it clean,

* *Batting*, a provincial phrafe for thrafhing.

it

it is well; and if foul, and you find fault, the anfwer is " Get fome body elfe ;" and he moftly quits your employ.

———

Mr. Johnfon, of Wilmflow, whofe machine is not fo large, and confequently not fo powerful as Mr. Harper's, winnows corn, and grinds oats or beans at the fame time.

Mr. Johnfon has the friction-wheels Mr. Harper means to adopt.

The following is a defcription of the Reverend Croxton Johnfon, rector of Wilmflow's, thrafhing-machine :

Diameter of the horfe-wheel—twelve feet.

D° tumbling fhaft nut—two feet.

D° of iron-wheel at the end of the tumbling fhaft within the barn—three feet eight inches.

D° of the nut—one foot two inches.

D° of the drum—four feet two inches.

D° of the pully fixed on the thrafhing cylinder—fix inches.

The diameter of the thrafhing cylinder—two feet.

The diameter of the feeding rollers—two feet fix inches.

The iron wheel on the feeding roller—two feet fix inches.

The nut on the thrafhing-cylinder—five inches.

The tumbling fhaft turns the iron wheel within the barn, which iron wheel turns the nut, on the axis of which the drum is ftaked.—The machine is turned by a ftrap from the drum round a pully fixed on the end of the axis of the thrafhing cylinder.

A nut is ftaked on the fame axis, which turns a wheel ftaked on the lower feeding roller.

Under the thrafhing cylinder there is a femicircular board, on which the ears of corn are beat by the fix beaters of the cylinder.

The fpace occupied by the thrafhing-machine is fix feet by three feet eight inches,"

A fwing-harrow has been lately introduced, and feems coming much into vogue.

A nut

Hurdles, of an improved conſtruction, merit notice. They are faſtened by a wooden pin, through a ſtrong piece of oak, in a manner ſo as to be looſened and removed with leſs trouble, and leſs injury to the hurdle, than the old forms. Theſe were obſerved at Mr. Bayley's, of Hope.

Winnowing machines, of an improved conſtruction, have been introduced, and gain ground: they diſpatch work briſkly, and ſave the chaff.

This machine cannot be too generally recommended, nor ſpoken of in too warm terms. It is admirable for its expeditious and neat manner of winnowing; and in the cleanſing of ſmutty wheat it is invaluable. I will venture to predict, that in a few years it will be in the hands of every farmer in the kingdom, who plows 20 acres of land. It is made at Aſhton, near Newton. Price £. 5. 5s.

A machine for cleaning corn from ſmall ſtones, or earth, of which foreign cargoes are, ſometimes, too full, and invented by Mr. Whiteſide, of Lancaſter, ſhould not be unnoticed: as alſo an invention of the ſame ingenious perſon, for opening, ſhutting, and bolting the doors of granaries, or corn-rooms. He imagined, that treading upon, and walking over the corn, to ſhut the door, or window, which admitted the air, was injurious; he therefore contrived a bolt, which opens the window, and ſhuts it again, by only pulling a cord, which runs upon a pully, and communicates with the ſhutter. The contrivance has both ſimplicity, neatneſs, and ſecurity.

CHAPTER

CHAPTER VI.

OF INCLOSING.

THE ground work of improvement muft be a general inclofure bill *.

Mr. Elkington's principles of draining, if publicly known, would be of moft material fervice in all new inclofures; as from his knowledge in difcovering fprings, he could place the fences in fuch fituations, that each ditch would anfwer the double end of fence and effectual drain, which hitherto, from want of his knowledge, has always required both operations.

There are but few open, or common fields, at this time re-maining; the inconvenience attending which, whilft they were in that ftate, have caufed great exertions to accomplifh a divi-fion, in order that every individual might cultivate his own lands, according to his own method; and that lots of a few acres, in many places divided into fmall portions, and again fe-parated at different diftances, might be brought together into one point.

The inclofures, or fields, are in general very fmall; fo much fo, as to caufe great lofs of ground from their number, and from the fpace occupied by the hedges, banks, and ditches. This great number of fences too, prevents the air from freely circulating, by which the crops, both of corn and grafs, are deprived of the falutary effects of the fun and air, and, after the grain is reaped, the procefs of drying, healing, &c. is ma-terially delayed.

There are objections to very large fields, which are, the cattle lying in them are more expofed to the weather, and can-not lie out fo long in the autumnal months. In ftormy wea-ther, nature and felf-prefervation teach them to feek the beft fhelter. And with refpect to corn land without fhelter, ex-

* This idea feems fo generally to prevail, that I am fure it cannot fail of being one of the objects that will be recommended to the legiflature by the honorable board.—*W. Dann.*

pofed

poſed to the ſea (as a great part of this county is) the winds blowing from thence, ſtrongly impregnated with ſalt water, would greatly injure the young crops, and ſhake the growing corn ready for the ſickle, as we but too often have experienced.

Beſides, the banks are.full of weeds, which often remain unmoleſted, the ſeeds of which are diſperſed by the winds over the adjoining fields, to their no ſmall injury. The hedge-rows are but too frequently neglected, and permitted to ſpread their branches upon the lands. Plaſhing is almoſt all neglected, except only by a few ſpirited gentlemen. Many hedges ſeem faſt upon the decline, and muſt in a little time be renewed. Durable as hedge-timber may be, a length of years brings on old age, and, at laſt, decay. The newly-planted hedges are chiefly of thorn, which muſt be the beſt, without intermixture, as formerly, of hazle, alder, willow, holly, &c. * The hedge-rows, which the ſurveyor has planted, are thorns upon the plane, without either ditch or bank, ſecured by rails, till grown up, and then trimmed, ſo as to meet in a narrow point at the top. Theſe fences are neat and ſecure ; and are preferable to hedges, cut ſquare at top, which are generally thin in the bottom. Many fences, particularly in the northern parts, are made of ſtone, ſome from quarries, and ſome of pebbles. Buildings are frequently erected with the latter, uncouth and misſhapen as they may appear.

The banks being liable to weeds (ſo are all other lands that are turned up) can be no objection to making them, as a mower would clean a new-made bank in a ſhort time; and this mowing would not be requiſite above a year or two (if care is taken that the banks be made of ſoil only) by that time the bank graſſes over, and becomes good herbage.—Railing may anſwer gentlemen's purpoſes, but it is too expenſive for common practice, or farmers.—I have always conſidered thorn

---

* The young ſhoots of the new-planted thorn are liable to great injury, if not well ſecured from cattle, who eagerly nip the tender ſprouts, and greatly injure the ſtem. The hair from a raw hide, with all the impurities adhering, if laid in ſmall quantities, near the roots of the thorn, have been found ſufficient ſecurity from the teeth of cattle. The cows will not approach near hedges thus defended.

quicks as beft adapted to fences, as they are both durable, and make the beft fence: and as the common method of planting them is liable to many objections, fhall give a method I have practifed with fuccefs; and for the better defcription have made the figure below, where A B C D reprefent the ditch; E F the water-table, upon which the quicks are to be planted; L F G H the cop, to defend the quicks.

The land, B L, is firft dug up one fpade deep, and near two in breadth; care being taken not to come too near the fide of the ditch, B, for fear of its falling in: then the bank is begun by taking out of the furface of the ditch, A B, one fpade deep, and placing them at E, green fide in: then clean foil may be got to form the water-table, E F, which ought to be eight or ten inches above the furface A K; upon the middle of which

the quicks fhould be planted, with fome old rotten dung, five in a yard: then proceed with the remaining part of the ditch, to form the bank E G H. Nothing more is ne-

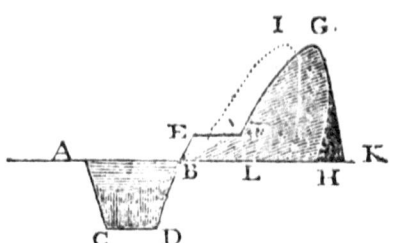

ceffary, except a dead hedge at the top of the bank, and keeping the quicks clean from weeds two or three years, when they will have grown fo as to make a tolerably good fence. The advantage of making a broad water-table is, that the quicks receive the benefit of the rains, by every fhower penetrating to the roots, which greatly invigorates the young plants, infomuch that as good a hedge will be formed in three years as would have required ten or twelve by the old method, which is, by making a narrow water-table, as from E to the dotted line, and planting the quicks up to the dotted line (E I forming the front of the bank), the point E, by frofty weather, and weeding, &c. falling away, makes it one continued flope as B I, which acts like the eaves of a houfe, carries off th rains which fhould refrefh and invigorate the young plants, and is the reafon fo many new planted hedges mifcarry.

Another method of planting quicks, which I think not a

bad

bad one, is planting them at the top of the bank; but the bank in this cafe fhould not be made fo high, and broader at the top; and the bank and ditch making one continued flope, it is neceffary to take care that nothing but good foil be put into the bank, and the quicks dunged as before.

In the firft method it is fometimes a practice to dig the bank floping down (after the quicks are become a fufficient fence) to dung and fet it with potatoes; and a great part of it afterwards is carted to the meadow or pafture land, as manure, but fo much of the bank is left as to prevent the cattle from treading upon the roots of the quicks.

Inclofures in this country are, for the moft part, infinitely too fmall; from two to ten ftat. acres may be the medium fize.—This unneceffary multiplicity of fences caufes much ufelefs expence, deftroys vaft tracts of excellent land, harbours vermin and nuifances of all defcriptions, and in reality rather prevents than facilitates the efflux of redundant moifture; the hedge-banks unreafonably high, devoid in general of timber trees, and even in a great meafure of every thing that can form a fence. Where white thorn is planted it is generally placed upon a narrow ledge or " water table," and an immenfe mound of earth erected behind it, which in procefs of time drives the plants into the ditch below; if the ditch be marle, the action of the froft fpeedily undermines the heavy bank, and by this means are produced thofe irregular and unfightly divifions equally ill adapted to the purpofes of utility or ornament.

Many parts of the county are very flat and wet, fuch confequently require good ditches; from this has arifen the multiplicity of ditches, and from that caufe neglect.

If fuch ditches as are abfolutely neceffary were properly attended to, and care taken to fecure a proper communication with the brooks and rivers, nine-tenths of the prefent fences would be unneceffary, as under-draining would amply provide againft all defects.—At prefent, moft of the ditches are nearly navigable, and no attention paid to gain an outfall, fo that they are full of putrid water, and are a perfect nuifance. Many hundred acres of excellent arable and pafture land are facrificed to this ftupid rage for fmall inclofures. I believe they may be of
fervice

fervice in keeping up the breed of water-rats, but I know of no other advantage to fociety arifing from them.

In every anfwer received to the queftion, Whether inclofures have increafed or decreafed population? the reply has univerfally been—increafed.

And how can the fact be otherwife upon rational grounds? In confequence of inclofures and divifion, every occupier has unqueftionably the means of cultivating his lands to the beft advantage to himfelf; but he cannot effect this without affording advantages to the public at large. Superior cultivation requires more labour, which requires a greater quantity of hands. The lands yield increafed returns; and produce both means to increafe population, and give food to the increafe upon better terms.

As to increafe of rent, the lands formerly in common fields but now divided, have doubled, in many inftances trebled, their rents immediately to the landlords; have yielded greater profit to the tenant; and have afforded more means of fubfiftence to the public.

The commons, or uncultivated lands, which heretofore have not yielded profit either to the proprietor or public; have increafed in their value from—nothing, if ftarving a few geefe, lean kine, producing—weeds, heath, &c. can, with propriety, be called nothing, or, to give fome better *ratio*, from one to thirty *per cent* [a]. In many inftances, the cultivated waftes have proved more fertile and productive than the old lands; if, therefore, the foregoing premifes be well founded, the public have gained 30 *per cent.* of additional employment and additional produce, by the improvement of waftes and commons; and the proprietor has gained, not indeed 30 *per cent.* for he has the expence of the improvement firft to deduct; but, on a moderate calculation, an addition of 10 or £.15 *per cent.* to his eftate, on the capital advanced.

Mr. Wilkinfon's improved mofs land, was, before draining, worth from 7 to 10 *s. per* acre, is now worth from 4 *l.* to 5 *l. per* acre of the large meafure.

Warbreck Moor, in Walton, inclofed in 1761, was not worth 1 *s. per* acre in its uncultivated ftate, is now well worth 30 *s. per* acre. After the inclofure act was obtained, and a divifion made the fee fimple of feveral lots was fold after the rate of 3 *l. per* acre, large meafure.

The

The furveyor has been informed of only one inftance where an attempt to improve wafte lands has failed.—Elland Moor, near Lancafter, notwithftanding lime has been laid on, and the ground treated according to the ufual cuftom of improving waftes; yet, after a few crops taken, feems verging back towards its original ftate of poverty.   But this was owing to its being overplowed at firft, and it is now coming round again.

Hedges.— When hedges grow thin at the bottom, Mr. Harper has the following practice.   He cuts the wood very low, leaving the young and vigorous fhoots; after cutting away the old wood, he takes a hand-faw, and cuts away again that part of the old ftump, fo far as was fhaken by the hatchet in the firft feparation, and faws the top level, fo that the water may not remain.   By this practice, he fays, the fhoots will grow ftronger, and more in number, in one year, than they would by the common practice in three years.   When the fhoots are half a yard, or two feet long, he bends the young fhoots down, and, where room permits, makes a hole in the bank with a fhovel, in which the fhoots are clofely tied down with hooked fticks, and covered up again with earth; when thefe young branches, with a little nurfing, will, by taking root afrefh, form a new hedge.

Gates—Are in general made of oak, and with five bars; fome are inferior, and made up of the materials, which the eftate may produce, and put flightly together, at a fmall expence. There are gates alfo made of deal, and painted.   In fome places rails only to hang up, and take down are made to fuffice.   About gentlemen's feats, there are gates frequently of fuperior conftruction, and made in different forms.

CHAPTER

CHAPTER VII.

ARABLE LANDS.

SECT. 1.—*Tillage.*

THE ploughmen are fuppofed to be as complete workmen as in any part of the kingdom, the tillage of their lands being in general performed in a mafterly manner. There is no general rule without fome exceptions; and it muft be acknowledged that fpecimens of great flovenlinefs might be produced.

In laying down either for corn or pafture, particular care fhould be taken to throw the butts or ridges as near north and fouth as poffible. Inftances are known where they have been laid eaft and weft, and in large round lands, when the fouth fide has yielded double.

SECT. 2.—*Fallowing.*

IN fome intelligent letters, which the furveyor has received, in anfwer to the queries which have been circulated by the Board, very oppofite opinions have been held upon this fubject. According to fome, fallowing is too little, according to others, it is too much practifed.

From what has been faid before, it is evident, that fallowing is here underftood, as preparatory for wheat. The tenant being generally under a covenant, reftraining him from fowing wheat upon clover, * whence a crop has been, the fame year, previoufly gathered, or from a bean ftubble, &c. as a practice tending to exhauft, and rendering the ground foul, which, by way of reproach, is called ftubbling †. Upon the fyftem of

---

* To reftrain a tenant from fowing wheat upon clover-lay is the greateft abfurdity that can be, becaufe it is the beft tilth known on moft foils.—*Mr. Boys, of Kent.*

† In many parts of the county the ground is not broken up for a *fallow* till fpring feedings are over: the abfurdity and want of fuccefs arifing from fuch management, forms no argument againft fallowing upon proper principles. But there is a very ftrong argument againft fallowing at all in this country, which is, that nine-tenths of the county are a fandy loam, capable of producing uncommon crops of turnips, cabbages, potatoes, &c. &c. and that manure is every where to be had in plenty.—What is the ufe of a fallow upon fuch land?

H

green crops preceding wheat, by way of faving one year's rent, and the labour of fallowing, the potatoe crop fhould feem to claim a fuperiority; both from the dung given, and the clean ftate into which, under good management, the land is brought. Yet the neateft farmers feem at prefent not very partial to this mode of agriculture. They fay the fucceeding crop of wheat is more feeble and worfe fed; and the bad effects of thefe two, potatoes and wheat in fucceffion, are evident upon fucceffive crops for years afterwards.

Fallowing may undoubtedly be avoided, by a well conducted variation of crops; and fo that the grounds may be kept clean by a full produce. Since every plant and fruit only extracts from the foil fuch fubftances as are required for its peculiar nature, and rejects thofe which are proper for the nutrition of others, e. g. the pungent tafte of an onion muft require very different juices to thofe which are neceffary to yield the mild flavour of a potatoe. Therefore by a well-regulated change the earth may be faid to have repofe equal to the fallow year. Gardens have been fucceffively cropped, or it may be doubly cropped, every year for a fucceffion of fifty years together.

Others affert that fallowing is a good preparatory for wheat, if the land is inclined to clay or a ftrong nature; but if, on the other hand, the land is of a light nature, fuch as a hazle loam, fand, or gravel, in that cafe fallowing is not preparatory to wheat, but would endeavour to come at clover roots to fow wheat upon. There is much fallowing for wheat in this county upon light fands, which is the height of folly.—A clover root is the beft preparation for wheat upon fuch foils. Beans are always fown in this county broadcaft, generally when the land is foul fowed with previous crops of corn.—Can the landholders be cenfured for difcouraging a fucceffion of wheat after beans neither hoed nor weeded?

Mr. Henry Harper prefers potatoe land for wheat, to that from whence a crop of turnips has been taken. But he prefers a good fummer fallow to either preparation. The grain produced from the fallows being of fuperior quality, and not fo fubject to blights.

The

The fallowing, as it is fometimes practifed, is not performed in the neateft manner. The lands not being broken up, till too late in the feafon to partake of the influence of the frofts, and to furnifh a proper opportunity for the crofs cutting, or ftirring (as it is called), this work being the grand operation ; and it is requifite, that a dry feafon be caught whilft the land lies open, and a large furface expofed to the influence of the fun and air. But if greater attention were paid to the turnip culture, with proper hoeing and dreffing, fallows would become lefs neceffary; and, according to the prefent advanced rent of land, they are too expenfive for the tenant.

Sect. 3.—*Rotation of Crops.*

Oats are univerfally fown towards the north-eaft and fouth-eaft of Prefton for years together, except this chain be broken occafionally by a crop of potatoes, and afterwards wheat, or wheat on a fummer fallow. In the Filde, which, from its fertility, has been called the Granary of the county, the foil has been ftill worfe abufed. Certain fields have been kept under cultivation, it is afferted, for more than a century, without intermiffion, under the following rotation. After marle, 2 or 3 years oats; beans or barley, each one year. If beans, barley the year after; but, if barley, then beans, and this alternate change of beans and barley continued for a few years. The eighth year from the firft marling is generally reckoned a period, from which the land is upon the decline, and a complete fummer fallow is given, and fome *till** (as it is called) is added, upon which, wheat; after the wheat, two crops are taken, one either of oats, beans, or barley; and then another fallow, with the addition of *till,* and two more crops of grain, above fpecified; and the practice, it is faid, may be advantageoufly followed for the fpace of 20 years, but is often continued much longer. Upon fuch courfes it is unneceffary to dwell longer, as they can afford neither pleafure, nor inftruction, to the experienced cultivator.

---

* A compoft of earth and lime mixed. Yard dung, and fea-mufcles, have been ufed, but this laft article is not found in fufficient quantities, nor is it durable.

I fhall

I fhail proceed to fome other practices, which are followed in other parts of the county *.—Oats, fallow, and next year wheat; if for barley or oats, the land to be manured and laid down with red clover and grafs feeds.   2. Oats or barley with dung, if rich, another crop, barley, then fallow for wheat; afterwards barley, with dung, and then laid down with clover and hay feeds.   3. Wheat, with one furrow, barley, with dung; fallow, for wheat; barley, or oats, laid down with clover, &c.   4. Potatoes, wheat, barley, with dung, and laid down.   5. After wheat, fallow for turnips, with dung, laid down with well dreffed hay feeds, from the cleaneft and beft meadow lands, with a mixture of white clover.   6. Early potatoes, after which a crop of turnips, then wheat or barley.   7. Early potatoes, and fown with grafs feeds, and white clover, without any corn; the hay fuffered to ftand, till the feeds become ripe, to drop and fill up vacancies, the ground well dunged after the firft crop of hay.   8. If the land be full of rufhes, by only taking a fingle crop of oats in the following manner; by plowing one furrow with a good dreffing of dung, harrowed in, upon which the crop of oats, with grafs feed only: by which the rufhes are deftroyed, but the grafs roots are preferved, and the grafs meliorated by expofing the foïl to the air and fun, by turning it once over.

Nothing can be fo barbarous as the rotation of crops in this diftrict; if that can be denominated a fyftem of rotation which depends merely upon the caprice of the cultivator, or upon what he thinks the land is capable of producing for the moment. Near Prefton, the general plan is to grow as many crops of oats in fucceffion as the land can produce, then fallow for wheat, by way of cleanfing the land; and then oats again, while oats can be produced; after which weeds and rufhes, 'till reft again

---

* In a note, however, take the following wretched rotation, which has been frequently practifed, and which has reduced both farmer and foil to an equality of poverty.

An old poor pafture broke up without being previoufly marled : 1. oats, 2. fallow, 3. wheat, 4. oats, 5. vetches and wheat, 6. oats, 7. fallow, 8. wheat, and this laft crop probably footed ; in which ftate the land is fuffered to remain till again reftored by that Power which can not only reftore, but create.

produces

produces grafs.—An occafional crop of potatoes fometimes intervenes, after which wheat,

Wherever green crops, fuch as turnips or cabbages, have been attempted, they have yielded immenfe returns, and fuch as ought to encourage the cultivation of thefe ufeful plants. But the application of them to fheep has been little attempted, though there is every reafon to imagine the introduction of fheep would be attended with the happieft effects,

Diftribution of Crops of a Field of Three Acres, of Eight Yards to the Rod, for the Years 1791, 1792, and 1793; fhewing the Amount of all Out-goings, Rent, &c. and the Quantity of different Produce of each Year, and the Amount it fold for : firft Beans and Turnips, fecond Vetches, third Wheat,

By Mr. Henry Harper.

	£.	s.	d.
Rent for three years, at £. 4. per acre - -	36	0	0
Taxes for ditto, church, king, poor and conftable, highways - - - - - - - -	4	19	0
Manure for ditto, 60 tons per acre, at 15 s. per ton - - - - - - - - -	45	0	0
Cartage and putting on the land, at 3 s. 6 d. for every ton and a half, and fpreading in the drills	21	5	0
Seed beans 2 quarters, at 36 s. per - - -	3	12	0
Twice ploughing for ditto - - - -	4	4	0
Drilling and covering - - - -	1	1	0
Sowing beans in the drills - - - -	0	10	6
Horfe-hoeing twice - - - - -	1	1	0
Hand weeding - - - - -	0	2	6
Reaping beans - - - - -	1	16	0
Cartage home - - - - -	0	14	0
Pitching to the cart - - - -	0	3	4
Thrafhing beans - - - - - -	2	5	0
Cleaning for market - - - -	0	3	0
Carting to market - - - - -	0	4	0

Seed

	£.	s.	d.
Seed turnips - - - - - -	0	6	0
Horfe hoeing for ditto - - - - - -	0	10	6
Sowing turnips broadcaft - - - -	0	3	0
Covering feed with horfe machine - - -	0	10	6
Hand-weeding turnips where too thick in the beans - - - - - - - -	0	2	6
Drawing 400 bufhels of turnips - - -	0	8	0
Cartage home - - - - - -	0	8	0
Seed vetches - - - - - -	3	3	0
Twice ploughing for ditto - - - -	4	4	0
Sowing broadcaft - - - - -	0	3	0
Harrowing ditto - - - - -	0	15	0
Mowing ditto - - - - - - -	0	18	0
Making them for hay - - - -	0	18	0
Cartage home - - - - - -	0	18	0
Pitching to the cart - - - - -	0	6	0
Seed wheat 1 quarter and half, at £.2. 12s. 6d. per - - - - - - -	3	18	9
Four times ploughing for ditto - - -	8	8	0
Sowing broadcaft - - - - -	0	3	0
Reaping ditto - - - -	1	16	0
Thrafhing ditto - - - -	3	3	0
Cleaning - - - - -	0	4	6
Cartage home from field - - -	0	18	0
Pitching to cart - - - - - -	0	4	0
Carting to market - - - - -	0	5	0
Tythe for vetches, at 6s. per acre, all others taken in kind - - - -	0	18	0
The average profit of three years crops, after all expences are deducted, at the full cofts - -	30	10	4
	187	2	5
By three acres of beans, at 8½ quarters per acre, 36s. per - - - -	45	18	0
By ftraw from ditto, 80 thrave, at 1s. per -	4	0	0

By

By 400 bufhels of turnips, at 8 *d.* per bufhel, at 　£.　*s.*　*d.*
　90 lb. per bufhel　-　-　　-　　-　11　13　8
By vetches, 500 ftone per acre, at 8 *d.* per
　ftone, £. 16. 13 *s.* 4 *d.* per acre　　-　　-　50　0　0
By three acres of wheat, at 8 ½ quarters per
　acre, at £. 2. 12 *s.* 6 *d.* per quarter　　-　-　66　18　9
By ftraw from ditto, 22 thrave per acre, at
　2 *s.* 6 *d.* per thrave £. 2. 15 *s.* per acre　　-　8　5　0
By light wheat, 2 bufhels, at 3 *s.* 6 *d.* per　-　-　0　7　0

　　　　　　　　　　　　　　£. 187　2　5

　This was a poor run-out field, that had been ploughed, &c. for near a century, and without any improvement at all only fince it came into my hands.　For the firft, I marled it, at 8 rod to the acre, of 64 cubical yards to the rod, nine years fince which it has never done any great things : then the laft three years before defcribed, for which I muft give the merits of the crops to drilling and hoeing, by keeping the land clean ; and, in the firft place, producing three bufhels for two if they had been fowed broadcaft, this I know by dear-bought experience : and, in the next place, 400 bufhels of turnips, which were worth £. 11. 13 *s.* 8 *d.* and all the ex-pences of feed, hoeing, drawing, and carting home, is only a difcount of £. 2. 8 *s.* 6 *d.* which is accounted for in the out-goings before mentioned, and the crop of beans even as luxu-riant and as proveable as where there were no turnips, and the land left in better condition.

　The manure was all put on the land for the beans and tur-nips ; and after producing the other two crops, vetches and wheat, the land appears to be left in a deal better condition than before it produced thefe crops, and if the rotation of the fame crops had been continued, 40 tons per acre would have anfwered as well as the 60 tons had done for the before-mentioned crops, which would have been a faving of £. 22, which is an object worthy of notice ; but as through conve

　　　　　　　　　　　　　　　　　nience

nience have changed the rotation of the fame crops to another old ploughed field.

One acre of the three had been pared and burned about thirty years fince, which, from that time, after producing two crops, never gave fcarcely the feed again, and never would give any grafs; and fince it has been in my hands, if the crops were ever fo luxuriant in ftraw, the corn was never fo well fed as the other part of the field, nor fo much in quantity according to breadth; only thefe laft three years it appears to have come round according to the other land.

The foil is a black fandy loam of a regular depth of about ten or eleven inches, under which there is a hard pellet of four or five inches thick, which is commonly called red ore, and under that, good marle fix or feven yards deep.

The field is called the Fernel, lying up to the townfhip of Orrel, about half a mile north of the Greavehoufe, in the townfhip of Bootle.

SECT. 4.—*Of the Crops commonly cultivated.*

THE grain principally cultivated is oats, which, when ground to meal, is the food of the labouring clafs, particularly in the northern and eaftern borders of the county; it is made into bread-cakes, of which there are various kinds, prepared by fermentation with four leaven; others without leaven, and rolled very thin; alfo water, boiled and thickened with meal into porridge; and this, eaten with fweet * or butter-milk. Small-beer fweetened with treacle, or treacle only, was in many families, about forty years ago, both the breakfaft and fupper meal. The general ufe of tea, efpecially among the females, has leffened the ufe of meal at breakfaft; and the influx of wealth has induced numbers to indulge, upon many occafions, with the wheaten loaf. Notwithftanding the

* Sweet milk is a provincial term, in contradiftinction to the buttermilk, which in this country is four, and therefore fometimes called four milk.

confumption

confumption of oat-meal is not fo general at prefent as it was formerly; yet the quantity ftill ufed is very confiderable; and the growth of oats is greater in proportion, than that of any other grain. There are fome excellent wheat lands, *e. g.* Low Furnefs, the low lands near the fhore beyond Lancafter, the Filde, and the S. W. part of Lancafhire; but wheat does not fucceed well, when bordering upon the moor lands; neither does barley, which feems, of the two, more delicate in foil, and there is a greater diminution in the cultivation of this grain, than of either wheat or oats. Beans, peas, &c. are alfo cultivated, but feldom drilled; a fmall quantity of buck-wheat alfo, for the ufe of poultry, or to be ploughed in previous to a crop of wheat, but very little rye is at prefent fown.

The tartarian, or reed oat, for fome years paft, has been the favourite fpecies of this grain, in the neighbourhood of Liverpool. Its produce is great, but the grain inferior, and not yielding an equal proportion of meal with the early or Dutch oat. The ftraw is luxuriant, and feems well adapted to lands exhaufted under bad management; nor is the grain fo liable to be fhaken out with the north-weft gales, to which this county is expofed, as the other fort.

Potatoes.—Lancafhire was the firft county in this kingdom in which the potatoe was grown: and as it is able at this day to boaft a fuperior cultivation in that important article, in which it ftill ftands unrivalled, it may be requifite to defcend to particulars in regard to the management of that crop: 1. A fward, or frefh lay, is defirable, but not always to be obtained. Good crops have been frequently raifed from lands exhaufted. The ground being previoufly cleaned by ploughings, and planted (if the ground can be got into condition) in April, in drills *

* I am confident that this method of planting either the early or late potatoe, is not fo productive as that of fetting them in beds of five feet wide, and covering them, when the fhoots begin to appear, with mould dug from a trench between the beds. This is the general mode in the neighbourhood of Frodfham, in Chefhire, where the planters of this moft valuable root have tried all poffible methods, for many years, and are generally allowed to produce a greater crop on a given quantity of land, than any other people in the kingdom.—*T. Wright*

about 3 feet diſtance, and from 12 to 9 inches aſunder, in each drill, the ſets placed immediately upon long dung from the yard, &c.; but dung from the great towns produces a wonderful effect upon lands not formerly accuſtomed to that article; and it is ſuppeſed, will generally enrich twice as far, with equal effect, as the manure formerly uſed from the farm-yard, &c. This is experienced in the lands bordering upon the canals. The great quantity of corn, and different kinds of provender, given to cattle kept in towns, muſt tend to enrich the quality of the dung, which depends upon the food taken, whether of man or beaſt.

2. Although April be the prime ſeaſon for producing a crop of good potatoes for the table, becauſe this vegetable requires a certain portion of time, to acquire that degree of maturity, which renders it peculiarly mellow and farinaceous, yet it is frequently planted as late as May, or even June; and yet produces abundant crops, but not of the ſame matured quality, as thoſe planted at a more early ſeaſon.

3. The apprehenſion of froſts (by which, if the tops are caught, after breaking the ſurface, they pine and ſicken, and the hopes of the huſbandman are blaſted,) ſometimes operate againſt planting at this early ſeaſon; yet good planters riſque the chance of froſts, in order to obtain ſuperior quality.

4. The crops are kept clean from weeds by the plough, firſt by turning a furrow, left for that purpoſe, towards the young plants, as ſoon as they appear; and afterwards by turning the ſame furrow back from each ſide of the drill, and which is ſometimes, if very foul, harrowed by a ſmall triangular harrow, running through each drill. After the weeds have been ſo expoſed, the furrow is turned back again, and ſometimes the ſame plough, or a double-wriſted one, runs up each drill once

---

* The ſurveyor has made ſome experiments to aſcertain the beſt mode of cutting the ſets; for, if the potatoe be ſet whole, putréfaction does not always enſue; and a ſet of a large ſize, to a certain degree, is better than a ſmall one. The beſt method he has yet diſcovered, is taking off the ſprout, or noſe end, and the umbilical, or tail end, of the potatoe, leaving the middle entirely for the ſet; the worſt method of cutting the potatoe, as has been proved, is cutting the potatoe down the middle, from noſe to tail end; a practice but too common.

more;

more; befides the deftruction of weeds, the foil, by thefe operations, is loofened, expofed to the fun and air, which contributes greatly to improve the crop.

5. There are various kinds of feeds in ufe.—The ox-noble,
and clufter potatoe are planted for the cattle *; the pink-eye,
and a variety of others, with different kinds of kidney-pota-
toes for the table. The old winter red, as it is fometimes
called, ought to be mentioned for its peculiar goodnefs in the
fpring, when other kinds have loft their flavour; this potatoe
is then in its beft perfection; it has another quality, that of
never having been known to curl. There are alfo great varieties of early potatoes, and great attention is paid to raifing
new forts of the beft qualities from feeds, of what is called the
crabs, or apples, which grow upon the ftems. Mr. G. Green
obferves, that after many experiments he invariably found that
the watery potatoe (of which there are great varieties) have
fallen far fhort of the purpofe intended. That he has feveral
times, both through neceffity as well as for the fake of experiment, given the ox-noble to milch-cows, after the more farinaceous fort, *e. g.* the pink-eye, when the decreafe of both
milk and butter has been evident in a very fhort fpace, and the
beafts themfelves feemed much diffatisfied with the change.

6. Great attention is paid to changing the feed occafionally,
to prevent the curl †, the practice of obtaining frefh feed from
Scotland

---

* Of the clufter potatoe, the furveyor had an opportunity of viewing
the produce of a crop, lying upon the furface of the ground, after being juft
taken up, belonging to Colonel Mordaunt, of Halfall, in this county. He,
and an intelligent farmer, were both of opinion, that they never faw fo
large a crop; and yet, as they were informed, raifed without dung.

The clufter, or conglomerated, or Suffolk (for fo it is called by Mr.
Howard, who firft introduced it to notice) was cultivated in this county
25 years ago (*a*) from fets left by that gentleman with the Society for the
Promotion of Arts and Commerce.

Vide *Doffie's Memoirs*, vol. X. It has fince been produced from feed,
and, though much improved in fhape, retains the red colour and faccha-
rine tafte.

(*a*) By the Rev. Mr. Heathcote, rector of Walton, and Mr. William
Haliday, Anfield.

✢ The furveyor had the honour of receiving a premium from the Society for the Promotion of Arts and Commerce, in the year 1789, for a
letter on the Lancafhire method of preventing the curl. He has the plea-

Scotland (as was the cuftom a few years ago), is not now fo
frequent; a change from the mofs lands, and *vice verfa*, being
generally fufficient.   A change of land is alfo defirable, but
not always practicable: crops have been fuccefsfully taken, for
a fucceffion of years, from the fame land.

7. The produce of a crop is, on a medium, from 2 to 3 hundred meafures, or bufhels *, the ftatute acre.   The early potatoes are generally planted in beds, in rows about 8 inches
diftant, and the fets 4 or 5 inches feparate, becaufe the early
putatoes, being of a lefs fize, require a fmaller fpace; but the
advanced price thefe early crops obtain at market, render them a
profitable article to the cultivator †; who, befides reaping a
profit from this early produce, has his ground prepared for
another crop the fame feafon.   Mr. Waring, fteward to the
Earl of Derby, gave to Major Atherton the following account
of the produce of one acre of indifferent land in Knowfley.

1793—700 bufhels of potatoes, pink-eyes.

1794—92 bufhels of wheat, 70lb. to the bufhel, fold at
7 s. 6 d. *per* bufhel.   3 months later they would have fetched

fure to obferve, that the fact feems to be confirmed, from the general opinion and practice of the county ; nor did he obferve a fingle difeafed potatoe in the whole of his furvey—the crops were univerfally luxuriant.
This thought is improved upon by Mr. Thomas Wright, gardener to John
Fazakerley, Efq. Prefcot, who has fent fome favourite plants which had
caught the difeafe of curl, to the mofs lands, which change of lands he
expected would effect a cure.
   * By a bufhel of potatoes, is generally meant 90lb. before they are
cleaned.
   † Mr. Ecclefton took the furveyor to view a piece of ground, 30
perches (8 yards to the perch) the early potatoes raifed upon which had
been fold for 30 l. in the prefent year 1793 ; after which, a crop of turnips
had been grown, which, at 6 d. *per* bufhel, were worth 50 l. *per* acre ;
after which the fame land was to be cropped with wheat.

Remark on this Fact.

   " The grofs amount of the account of the potatoes appears to be great,
that of 20 s. *per* rod of 8 yards ; but if all expences of fets, and preparing
the land, and getting them up, and afterwards marketing them at the different markets, Liverpool, Manchefter, &c. were deducted, it is a query
but the outgoings would be confiderably more than the grofs amount given,
although the land muft be perfectly well prepared for the turnips ; but the
account given of the turnips, at the rate of 2000 bufhels of thirty-fix quarts
or ninety pounds *per* bufhel, is more by 800 bufhels *per* acre than ever I
knew or heard of for either large or fmall lot, either by hoeing, or any other
advantage to be taken."—*Mr. Harper.*

10 *s.* 6 *d. per* buſhel, cone wheat. Mr Waring ſays, the live crops were equal to the fee ſimple of the land. He is confident that *marle* would have *produced* 20 *buſhels more wheat.* The markets of Mancheſter, Oldham, Rochdale, and the neighbour-hood, are ſupplied with great quantities, not only from War-rington, but as far as from Rufford, Scarſbrick, &c.

Upon the ſame ground, from which a crop has already been taken, the early ſeed potatoes are in ſome places afterwards planted; which, after being got up about November, are immediately cut up into ſets, and preſerved in oat ſhells *, or ſaw-duſt, where they remain till March, when they are planted, after having had one ſpit taken off, and planted with another, of a length ſufficient to appear above ground in the ſpace of a week.

But the moſt approved method is, to cut the ſets, and put them on a room-floor, where a ſtrong current of air can be introduced at pleaſure, the ſets laid thinner, viz. about 2 lays in depth, and covered with the like materials, (ſhells or ſaw-duſt) about 2 inches thick: this ſcreens them from the winter froſts, and keeps them moderately warm, cauſing them to vegetate; but at the ſame time admits air to ſtrengthen them, and harden their ſhoots, which the cultivators improve by opening the doors and windows on every opportunity afforded by mild ſoft weather: they frequently examine them, and when the ſhoots are ſprung an inch and a half, or 2 inches, they carefully remove one half of their covering, with a wooden rake, or with the hands, taking care not to diſturb, or break, the ſhoots. Light is requiſite as well as air, to ſtrengthen and eſtabliſh the ſhoots; on which account a green-houſe has the advantage of a room, but a room anſwers very well with a good window or two in it, and if to the ſun ſtill better.—In this manner they ſuffer them to remain till the planting ſeaſon, giving them all the air poſſible by the doors and windows, when it can be done with ſafety from froſt: by this method the ſhoots at the top become green, leaves are ſprung, and are moderately hardy.

* Vulgarly called meal ſhudes.

They

They then plant them in rows, in the ufual method, by a fet-ting-ftick, and carefully rake up the cavities made by the fet-ting ftick; by this method they are enabled to bear a little froft without injury. The earlieft potatoe is the fuperfine white kidney*; from this fort, upon the fame ground, have been raifed 4 crops; having fets from the repofitory ready to put in as foon as the other were taken up; and a fifth crop is fome-times raifed from the fame lands, the fame year, of tranfplanted winter lettuce. The firft crop had the advantage of a cover-ing in frofty nights.

The above excellent information was communicated by J. Blundell, Ormfkirk, and has hitherto been known only amongft a very few farmers.

8. The manner of taking them up varies. The three-pronged fork is in general ufe—the foil turned over, the weeds picked out, the potatoes gathered and feparated, according to their fize, by the fame perfon. Another practice is, for a ftrong man to take a three-pronged fork, but crooked (the fame which is generally ufed to pull dung out of the cart) which he ftrikes down between every root, and pulls it over, laying the roots bare, which are taken up by two children that follow. Ano-ther practice is to turn a furrow from the potatoes, with a Ro-theram plough, and then with another plough, furnifhed only with a fhare, to turn up the potatoes, which are afterwards ga-thered.

After the potatoes are gathered, and fufficiently dried, they are put together in heaps, in the fhape of the roof of a building, covered clofely with ftraw, which fhould be drawn ftraight, and to meet from each fide in a point at the top, about fix inches in thicknefs, and then covered with mould, clofely compacted to-gether, by frequent applications of the fpade; after which Mr. Ecclefton makes holes in the mould, at the fides and tops of thefe repofitories, as deep as the ftraw, and about three yards diftant, to permit the air, which, he fays, vifibly arifes from the fermentation, to efcape: after the fermen-

* The early potatoe is a diftinct fpecies, of which there are yet great varieties.

tation

tation has ceafed, the holes are clofed to prevent the effects of frofts or rain.

9. The utility of the application of potatoes to feeding ftock, is fufficiently known, but not fufficiently practifed. Converting the produce into immediate cafh, by taking it to market, is a ftronger temptation than waiting the more tedious procefs of purchafing ftock, and fattening the cattle; but a fource of improvement to the land, and confequently of fuperior profit in the iffue, is by this means done away.

10. From the amazing quantities confumed by ftock, it may not be amifs to mention the manner of boiling, &c. which is almoft univerfally by fteam, in a large hamper, or tub, perforated at the bottom, and placed over the water: in this way they are readier for ufe than by being immerged in water; after which they are given either warm or cold, mixed with chaff, bran, hay feeds, barley, or oatmeal.

The method of boiling potatoes by fteam, has been adopted by fome for culinary purpofes as an improvement, thinking by this procefs they muft imbibe lefs water from their not being immerged in the fubftance. But immerfion in water caufes the difcharge of a certain matter, which the fteam alone is incapable of doing, and by detaining of which the flavour of this root is injured. The cottager underftands this kind of cookery: having poured off the water, he evaporates the moifture by replacing the veffel in which the potatoe was boiled, once more over the fire. Potatoes do not admit being put into a veffel of boiling water like greens. If America *, whence this choice vegetable was firft imported, had yielded nothing elfe to the refearches of the European, the prefent generation would have reafon to be thankful for the acquifition, and to be grateful to the planters in Lancafhire, for their fpirited attention to the cultivation of this excellent root.

* A note in a common-place book that I wrote feveral years ago, informs me, that John Hawkins, a dealer in flaves, got in 1565 the firft potatoes for fhip provifions from the inhabitants of Santa Fé, in New Spain; he introduced the root into Ireland, whence it was farther propagated through all the northern parts of Europe.

An old method of cooking potatoes.—Boil, and let them grow cold, then eat them, mixed with oil, vinegar, and pepper. *Parkinfon's Herbal.*

Turnips.—

Turnips.—It muſt be acknowledged, that turnips are not cultivated but on a very contracted ſcale*, and even then but ſeldom hoed; and yet there are not many articles more adapted to the ſoil and climate, there-being ſeldom a crop deſtroyed, or loſt-by the ſlug (or whatever that is which deſtroys the tender plant). The turnips find a ready market † if near a great town, whilſt the inferior crops generally pay well, if applied wholly to feeding cattle; and they leave the land in ſo clean a ſtate, as to be fit for moſt kinds of grain, and generally taken, by the beſt farmers, as a previous crop, to lay down to graſs or crops of clover.

Mr. Eccleſton not only ſows his turnips in drills, but every other ſeed, and was the firſt who introduced this vegetable into a ſyſtem of crops in his own neighbourhood.

Clover.—This ſort of graſs is cultivated generally with ſucceſs; being greatly preferred to the white hay, by thoſe who keep horſes in the great towns for the draught; containing, it is ſuppoſed, more nutriment. If opportunity offers, inſtead of ſending their horſes to graze upon a field, which is difficult to obtain, a lot of green clover is purchaſed, and brought in that ſtate to the conſumer, who ſoils his horſes in the ſtable for a few weeks in the year, and it acts both as food and phyſic, and enables them to ſtand work the better. Some few farmers keep their cart horſes in the houſe throughout the year, and ſoil them in ſummer entirely with clover.

The lands upon which clovers have been frequently grown, it is ſaid, do not yield ſuch plentiful crops as they did ſome years paſt; ſecond crops, in this northern climate, are ſeldom worth the riſque of being made into hay, and, beſides, are thought to exhauſt the lands, therefore are generally paſtured. But marle will always inſure clover; when it fails in this county it is the fault of the huſbandman, not the land.

---

* Turnips, to the amount of eight acres, were cultivated in the neighbourhood of Wrightington, by William Diconſon, Eſq. about 30 years ago: before this period none had been ſown but in the gardens.

† To raiſe an expenſive crop of turnips merely to *ſell*, may be good management with a gardener, but not with a farmer. The crop that is not conſumed upon the premiſes, I cannot allow to be a meliorating crop.— *T. W.*

Other

Other green crops.—Vetches are fometimes cultivated as a fmothering crop, and a preparation for wheat, but not very generally. Lucerne has been attempted, but at prefent not much, if at all, cultivated. Scotch cabbages have been planted, and good crops raifed, but not to any great extent. Carrots are fuccefsfully cultivated upon fandy loams, in the neighbourhood of Kirkby, Scarifbrick, Burfcough, Rufford, chiefly for the fupply of the Liverpool market, and fometimes purchafed to be given to horfes (particularly wind-broken)—They are generally fold about 2 s. 6 d. or 3 s. *per* cwt. and are reckoned a profitable crop on fuitable lands.

TIME OF SOWING.

Wheat feeding is from the middle of September to the end of October. Mr. Ecclefton, of Scarifbrick, fays, " The beft crop of winter wheat I have feen this year, or, indeed, ever recollect, was fown after a crop of potatoes, as late as the 20th of laft March. I mention this as an extraordinary fact."

The time of reaping wheat, from Auguft to September.

Beans are ufually fown early in March, and reaped in September.

Common oats in April. Early oats in May and June, and reaped in Auguft, September, and October. Barley is fown in April and May, and reaped in Auguft and September. Thefe are the general feafons.

But there are always exceptions to general rules; *e. g.* the prefent year the produce of feveral fields, both barley and oats, was not put into the barn, in the fouth-weft part of the county, the fecond week in November; and there was a certain field of barley in Toxteth Park, not cut the third week in November.

On the mofs lands, where paring and burning is practifed, both feed time and harveft is very late; owing to the uncertainty of the weather—if wet, the burning proceeds but flowly; the feed time is confequently retarded, and the crops are by thefe means fo late, as to become precarious from the advanced feafon, being frequently expofed to frofts and fnows. If the barley from the mofs lands be well houfed, it is in high eftimation: and obtains an advanced price from the farmer, who

K   prefers

prefers corn raifed upon thofe lands for his feed. Mr. Ecclefton fowed one year a field of barley about the middle of June,
which he houfed the following year, January 1. Barley is
generally fown too late in this county—much of it even in
June, but the greateft part in May—in the mofs lands, where
paring and burning is the preparation for this grain, this practice may have its foundation in neceffity; but the imitation is
abfurd on good barley lands.

HARVESTING.

The grain in this county has been ufually reaped by the
fickle, the quantity grown being but fmall, and the labourers
abundant. In the year 1794 feveral farmers however mowed
their corn, amongft whom was Mr. H. Harper, who fetched
the furveyor to fee his procefs, which was neat, and in the following manner.

The wheat was mown *in*, that is, thrown towards the ftand-
ing corn, immediately gathered and tied up into fheaves; the
fet was two mowers, two women gatherers, and one man
binder. The barley and oats were mown *out*, into fwathes,
and gathered at convenience. The advantages of this method
were, a faving of expence about 14 *d.* per acre, lefs danger of
the corn being fhook out of the ear, and gaining nearly one-
third more ftraw; no trifling confideration under feveral heads,
efpecially fince it does not appear that what ftubble is left in
the field is of the leaft fervice, in fome cafes evidently does
harm, *e. g.* to clover or other young graffes, by retaining
moifture through the winter, and ftarving the tender plants,
or injuring the hay when mown, and which, when wet, it has a
tendency to render putrid.

After the corn was gathered, the ground was gone over
with a rake, to collect what ftraggling ears might remain,
which are generally the heavieft, and of fuperior quality.

A wooden rake, with teeth about one inch longer than the
common hay-rake, was preferred to the drag-rake, and did
its work much neater—a woman could rake about two ftatute
acres per day.

The fcythe for cutting the corn had an addition of a bow,
made out of a piece of rod-iron, faftened into the pole, and ex-

6   tending

tending three inches over the fcythe-heel, from whence it rofe about nine inches in height and about two feet in length, and which formed a kind of cradle. The rod was fupported by an upright prop from the pole about the centre, and which was furthermore braced and kept tight by a ftring.

The Lancafhire method of fetting up corn, after being reaped, and whilft it continues in the field, may merit to be noticed; which if barley or oats, and in a greenifh ftate, is fet up in four ftandard fheaves only, with one cover called a hooder, that is, a large well-bound fheaf is felected and opened, with which the four ftandard fheaves, with the grain upper-moft; are covered, the grain of the hooder hanging downwards, but free from the ground. This fhape is provincially called a *pricket*. But the moft general method is, fix fheaves ftandards placed againft each other, fpread out in their butt ends, and clofed tight at their tops, when a couple of fheaves are opened, each about one half, clapped over each end of the ftandards, and meet with their butt ends together in the cen-tre, thus forming a roof or cover for the ftandards. This form is provincially called Hattocks, and their covers Riders.

PRODUCE.

Where the land is well cultivated, inftances of a great in-creafe might be given; but the general produce of the county cannot be ftated at more than 24 bufhels of wheat, 30 of bar-ley, and 40 of oats.

PRESERVATION AND MANUFACTURE OF CORN.

Corn is kept both in barns and ftacks: the laft is confider-ed to be preferable. There are mills belonging to the Free grammar-fchool at Manchefter, granted by Hugh Bexwick, Clerk, and Joanna Bexwick, widow, in 1524, where a great quantity of grain is manufactured. In the neighbourhood of Liverpool, moftly windmills, but there is one tide-mill lately erected there, which does confiderable work. The mills in general, are private property; and, except in few cafes, the te-nants are not bound to grind at particular mills. Where they are bound, great indulgences are granted.

SECT. 5.—*Crops not commonly cultivated.*

LIQUORICE,

Is not cultivated in this county, in any sufficient quantity, as an object of profit; although upon many grounds, it might flourish, and be worthy of attention.

The surveyor has a number of plants interspersed amongst other shrubs; when the root is wanted for decoctions, or other use in the family, a quantity is taken up, and it has been found to be as well-flavoured, rich, and juicy, as the Pontefract.

RHUBARB,

Also, has been planted in this way, a number of years, and the root cured and made use of; some pounds were lately presented to the Liverpool Dispensary. This plant, when in bloom, has a majestic appearance; its growth, at a certain period, a little before the seed appears, is amazing. The stem has grown, in length, three inches in twenty-four hours

The surveyor, has, at present, a most vigorous plantation. Having destroyed an old hedge planted upon a bank with a

---

* Growth of a rhubarb plant, N° 2, belonging to the surveyor, and measured by him in the year 1789.

	Morning.					Evening.		
	Feet.	Inches.	Tenths.					
May 11	3	6	5	-	-	3	7	9
12	3	9	8	-	-	3	10	5
13	4	2	5	-	-	4	3	4
14	4	5	9	-	-	5	0	0
15	4	10	8	-	-	5	5	5
16	5	2	8	-	-	5	7	8
17	5	6	4	-	-	5	10	1
18	5	8	2	-	-	6	1	0
19	5	10	3	-	-	6	2	7
20	6	1	2					
21	6	4	5					
22	6	7	5	-	-	6	7	8
23	7	0	1	-	-	7	1	6
24	7	3	1					
31	8	2	0					
June 5	8	8	0					

1795, a rhubarb plant of the surveyor's, which broke ground April 1st, was, June 15, 52 inches long; 16th June, in the space of 24 hours, grew 4 inches 6·10ths; from the 22d of the same month, 4 inches 9-10ths.— *Vide Gent. Mag. June* 1795.

ditch

ditch on one fide, a new thorn hedge was again planted where the old bank had formerly ftood, and the ditch filled up with rich earth, in which the plantation of rhubarb was made, fecured on one fide by the hedge, on the other by rails.

CHICORY.

Mr. Wakefield fpeaks highly of the heavy crops of chicory he has mown, from the fame land, and with which he has foiled his horfes, viz. ten horfes, the fpace of ten weeks at hard work, upon this plant, and without either hay or corn, from two ftatute-acres; and was cut 3 times in the feafon; firft time about the 20th of May; that which remained for feed grew to the fize of 8 or 9 feet high. The root of chicory is made ufe of as coffee in Germany, &c.

MADDER.

It was obferved, by an ingenious gentleman*, that madder, he imagined, might be fuccefsfully cultivated, and with advantage, upon mofs lands, fince the art of dying cottons a Turkey red has been difcovered, for which purpofe madder, in the root, is abfolutely neceffary. Madder, which previous to this difcovery was of little value, is now worth 50 s. per cwt.; and, if of prime quality, worth 120 s. per cwt. This root was attempted to be cultivated in this county fome years paft, under the encouragement of a premium, by the fociety for promotion of Arts and Commerce, but failed of fuccefs under the expenfive procefs of drying, by artificial heat, the difficulty of grinding, peeling off the bark, &c. But of late the fun has been found fufficiently powerful to cure it, and the grinding and peeling procefs is better underftood.

RUTA BAGA.

Mr. Taylor kept fix brood mares, and two young horfes 3 years old, upon the Swedifh turnip and ftraw, in a fold-yard.

---

* Leigh Phillips, Efq. Manchefter. A fpecimen of dying with madder of his own growth has been tranfmitted to the Board of Agriculture, and been viewed with much approbation.

They

They appear healthy, and in fair good condition, to each he gives half a bufhel a day.

The Ruta Baga, or Swedifh turnip, has ftood the fevere froft of 1794 and 1795, whilft the Englifh turnips of almoft every fpecies have fuffered, and upon the wet lands have been totally rotted and deftroyed. The tops of the Swedifh turnip it is true, have fhrunk; but the root ftands quite firm. This turnip is a valuable acquifition.

HEMP AND FLAX.

The culture, neither of hemp nor flax, was ever carried to any great extent in this county. It is proper to remark that a crop of hemp is fuppofed to be an excellent means of deftroying couch, let it be ever fo abundant.

Mr. Fazakerly obferves, that couch fhould always be deftroyed upon the land, by fmothering or withering; and if either carried off the land, or even burnt upon it, the ground is injured. He contends, from experience, that though the couch, whilft living, be injurious, yet it fhould never be taken from the lands whence produced, but the roots by fome means there deftroyed by putrefaction.

Chapter VIII.

GRASS.

Sect. 1.—*Natural Meadows and Pastures.*

ALTHOUGH there is a mixture of arable and grafs land, yet the latter muft greatly preponderate, and that to fuch a degree, that it has been frequently afferted, that the corn raifed in Lancafhire would not fupport the inhabitants more than three months in the year; fo that the eafieft way of obtaining corn, until the county is improved, is to purchafe it at other markets.

The lands in the immediate vicinity of the great towns are chiefly employed in pafturage; at a remoter diftance, in pafturage and meadow, immenfe quantities of hay being requifite for the number of horfes and cows kept therein. Near fome places, fuch as Bolton, befides the demand for lands under hay and grafs, a great number of acres are occupied as bleaching grounds; and throughout the whole of the county there are, in different places, many acres of rich land, covered with yarn, or cloth, under various operations.

Thefe feveral caufes have had a tendency to change the fyftem of the agriculture of the county, and to convert the arable grounds into grafs lands; and this fyftem of management feems yearly increafing, even in thofe parts which were formerly confidered as the great corn diftricts; fuch as that fertile foil under the denomination of the Filde, a tract of land from the north of the Ribble along the coaft as far as Cockerfands, to the turnpike road on the eaft.

At this period, (1795) the diminution of arable land is likely to become a ferious calamity to the nation at large. The converfion of arable land into grafs in this county may be imputed to feven caufes.—1ft. The enormous and immoderate wages to be obtained in the manufactories, which has wrefted the arm of induftry from the plough.—2d. The confequent encreafe of the poor rates, becaufe the manufactories do not fupport their own poor; and the manufacturers, if out of employment,

when

when fick, or infirm, or aged, are fupported by *taxes levied upon agriculture.*—3d. By all capitals being vefted in the working cotton inftead of raifing corn.—4th. To the very abfurd rotation of crops ufed throughout the county.—5th. To the barbarous cuftom of keeping the fame land too long under the plough.—6th. To an opinion, originating in the confequence of the two laft reafons, that grafs is more valuable than corn. Good grafs probably may, but not fuch grafs as is to be found through a great part of this diftrict.—And, 7th. To the exaction of tythes in kind.

Sect. 2.—*Artificial Graffes.*

The mode of laying down grafs for hay, is after having taken a few crops, cleaned and dunged the land, along with barley and oats, to fow the red clover, with the hay-feeds which fall off in feeding, which are collected; fometimes trefoil is added. Ray-grafs of late years has not been in much eftimation. Mr. Ecclefton, Mr. Wilkinfon, and Mr. Philips have each of late fown chicory or fuccory. The laft has already kept his coach-horfes three months upon this plant; they look well—the chicory is already fufficient to mow a fecond time—this plant caufes his horfes to ftale much.

Pafture lands are, in general, moft miferably laid down, they being in many places left to nature, to fupply the ground with whatever feeds remained in the earth, or came from other quarters, carried by the winds or other accidental caufes; and in the Filde particularly the lands have, on many occafions, been fo exhaufted by repeated plowings, that they are rendered incapable of yielding any ufeful herbage; feeds that have hitherto been tried upon thefe lands have fickened and died away, and fome have not even vegetated; and the furface remains covered with weeds of various kinds, for a fucceffion of years. White clover, and the cleaneft hay feeds, have been the beft fyftem of laying down paftures, hitherto practifed; but in attempting this, many of the farmers have been too inattentive to the choice of their feeds, which have been promifcuoufly collected as they dropped from the hay, without regard to the fpecies of

grafs,

grafs, the crops being free from docks or other fpontaneous weeds, which were permitted to grow. But the lands in general abound with varieties of natural grafles; and, if in tolerable condition, in a very little time will be covered with a good fward; among which, white clover, growing fpontaneoufly, is not unfrequent.

Inftead of the old method of laying down land in fmall ridges (called *butts* in Lancafhire) particularly in wet lands, of late the beft farmers have adopted the fize of fix or eight yards broad, with but fhallow intervals; if for mowing, the lands are in a better ftate for the fcythe; if for pafture, the cattle not fo liable to be overthrown in the deep drains. In very dry lands, which require no drains, the furface is laid as fmooth and even as can be effected; the whole being united into one plane, i: poffible; which not only renders the furface of the land more agreeable to the eye, but in every refpect of agricultural management fuperior. To prevent thefe butts being too high in the centre, the land is drawn out into breadths of half the fize of the intended butt, then a furrow is thrown together from each fide of the two, which are to be formed into one for the centre part.

Red clover is fown alfo, but not as a matrix for wheat, to which the land in fome places is adapted *. After two years crop of red clover, although hay feeds have been added, there is generally but a fcanty crop, the clover difappearing; and, unlefs an ample dreffing of manure be alfo given, the produce of hay feeds will be very fcanty; this mode of manuring is by good farmers frequently practifed. Some experiments have been made upon the *Alopecurus pratenfis* and *Feftuca pratenfis* with great fuccefs; as alfo the wild endive or chicory *(Cichorium intybus)*; but thefe trials are yet in their infancy, and the fcale but fmall. Trefoil, cinque-foil, rib-grafs, and rye-grafs, have been frequently fown, but in no great quantities, but this laft is feldom found to anfwer here. But the fame foil, in different feafons,

---

* If for pafture, red clover is omitted, white clover and feeds collected from the hay-lofts, are alone ufed. Some fields have been laid down to pafture, with grafs-feeds only, without any corn, and have been found to fucceed. There is faid to be an evident fuperiority in lands thus treated, although twenty years ago: but the experiments have been few. A gentleman at Bolton Moor has an excellent pafture the prefent year, with white clover, fown with vetches.

produces

produces different kinds of graffes, *e. g.* white clovers, which may probably arife from the application of different manures, or the feafons being more congenial to this or that fpecies of grafs.—The feeds muft be originally lodged in the earth, the great ftorehoufe where nature has depofited her treafures; for none have been fabricated, they have been only collected and felected by the induftrious cultivator to whom they offer their liberal aid. Tufts of knot grafs, which fcarcely any beaft will touch, have been removed by fpreading a little lime over them. Another fpecies of grafs has fucceeded this operation.

The great abundance of natural graffes in this country, fuperfede, in a great meafure, the neceffity of having recourfe to artificial ones. Sainfoin and lucern are unknown, or nearly fo. The land naturally produces white clover, efpecially when kept in high condition; the application of the root of red clover as a matrix for wheat, is fcarcely ever practifed, though admirably adapted to the lighter land of the county. It is however fown pretty generally when land is intended to be laid down to grafs; by this means the farmer obtains two very large crops of hay the firft year, but his land is much impoverifhed for the next two or three, as the clover difappears, and the natural graffes do not pufh forward, as the land has been generally haraffed by the previous crops of corn and clover. This refult is in fome meafure obviated by an ample dreffing of manure being given to the clover root, for manure is to be purchafed in this populous country in vaft quantities. Upon the whole, the manner of laying down land to grafs is by far the moft reprehenfible part of the management of this county. After land had been many years under tillage, the old plan of the country was to fallow for wheat, and leave the ftubble of little narrow wheat butts to produce whatever weeds and trumpery it might pleafe Heaven to fend: of late years, the ftubble has been well manured, and fown with barley and clover, and the refufe of the hay ricks. The manure, and the additional breadth of the barley butts, and the grafs feeds were an improvement; but in general this advantage was much diminifhed by the foulnefs introduced by the additional crop, the vigour and abundance of the couch grafs, and the foulnefs

x

of

of the hay-feeds. By the time the clover had been twice mown, the lands were in miferable condition, little but couch grafs and weeds to be feen: but reft from the plough, and the natural fertility of the foil, by degrees brought it into condition to be *ploughed again.* Such management has been productive of much lofs both to landlord and tenant, and is the reafon that gentlemen of property are fo defirous of having the *tillage of their tenants* fo much *reftricted.* We are, however, beginning to adopt a more enlightened method of laying down our lands: fallowing for turnips once or twice, if the land is very foul, and then fowing barley and well-dreffed hay feeds, from known good meadows, and white clover. Another method is, to manure land very well for early potatoes, which ought to be off the land in June, July, or Auguft at lateft, and fowing grafs feeds and white clover, without any corn; the hay ought to ftand until the hay feeds are pretty well ripened the fubfequent year, and the eddifh or after-grafs to be well manured as foon as the hay is carted off.

Sect. 3.—*Hay Harveft.*

In the management and curing of clover, which, from the quantity of moifture to be evaporated from the plant, before it be cured fufficiently to keep, is attended with confiderable difficulty, the following method has been practifed by Thomas Ecclefton, Efq. that fpirited gentleman fo frequently mentioned.

Hay, without doubt, cures fafter the more it is raked, as by this, more furface is expofed to the influence of the fun and air, by frequent turning and fhaking:—but, in my method, a very little labour, will fuffice when the weather is good. The only difficulty is to cure hay, fo as to preferve its nutritious juices, fcent, and other qualities, when the feafon is wet, and the grafs, through its different ftages, is repeatedly caught with fhowers.

Mr. Ecclefton's mode.—The clover is collected together into fmall fheaves, and kept ftraight; then twifted together, in the top part, to admit the fheaf to ftand upon its butt, or bottom end, when fpread out, in the fame manner that horfe-beans

have

have been frequently treated; and if thefe little bundles are not thrown down by the winds, they will refift more rain, if it fhould fall, than when lying on the furface of the ground; and if the weather be fine, having more furface expofed and open, the clover will cure the fafter.

In making hay-ftacks, befides a chimney * in the ftack, by a bafket placed in the middle, and drawn up by a cord, in order to fuffer the air, generated by heating, to efcape, and to prevent the ftack taking fire, as mentioned in the " Survey of Mid-" dlefex," Mr. Ecclefton cuts gutters in the ground, length-ways, and covers them acrofs in that place whereon a ftack is to be built. ,Through thefe trenches, in different directions, the outward air may enter, pafs through, then afcend the aperture left in the ftack; and this continued circulation takes away the generated heat or foul air, which, if confined together without any vent, might produce damage to the hay, or worfe effects; and, by thefe ufeful precautions, he is enabled to collect his hay together at a more early period, and in a more juicy ftate; by which good practice, time is faved, and the quality of the hay improved.

I have obferved ftacks of clover hay, made with layers of wheat ftraw, at certain diftances, from the bottom to the top, which I think a good method, particularly when it has had bad weather upon it, and was got in rather damp, as the damp heat is conveyed through it by means of the ftraw from one fide to the other, and a greater circulation of air might ftill be procured by a chimney in the centre being filled with ftraw.

Hay-barns have of late been erected in many places, ftand-ing upon pillars, and covered with flates; fometimes with a bottom boarded with planks, open in the joints, perforated with holes, and lying hollow a fpace above ground, to admit a free circulation of air all under the hay. Thefe buildings are ufe-ful, cheap, and bv their great convenience in bad weather, and

---

* " When hay is *properly prepared* to be put together in a ftack or rick, a chimney ought never to be made; it is a great evil, never to be adopted but when there is abfolute danger of the rick taking fire. Rather let an ox-feeder in North Wilts be confulted in the art of hay-making, than a farmer in Lancafhire."—*T. W.*

the

the great prefervation they afford to the hay, will foon repay the firft expence.

It is a good practice with hay in buildings, as foon as it is become folid enough to bear the knife, to cut a paffage round the walls, about half a yard in breadth. The hay which comes from the paffage thus cut, may be put on the top of the mow: by this method, a free circulation of air is obtained, and the tainted fmell which is contracted by the hay which lies up to the walls through the winter, is by this method prevented.

S F C T. 4.— *Feeding.*

THE common average of the beft lands, is one ftatute acre per cow, for the fummer's acre; but there are fome thoufands of acres that will fall greatly fhort, fome paftures being fo very poor as to require three, nay four times that breadth of land, not to feed, but barely keep alive, thofe poor beafts who have the hard fate to be doomed to the great labour of collecting their food fo fcantily and widely difperfed.

Lands under the higheft ftate of cultivation will keep and fatten even more than one beaft upon an acre.—The furveyor's fummer pafture in 1794, was about five ftatute acres, which plentifully fupplied five tolerably fized cows, two large horfes, and one of a fmaller fize, and feven pigs, regularly turned out to pafture twice every day, between their meals. Thefe pigs confumed a confiderable quantity of grafs, were admitted into the ftyes when their meals were prepared, and after having taken their reft, were regularly turned to pafture again. This feems no bad practice in the management of hogs; they grow faft, and their flefh is rendered remarkably fweet, which cleanlinefs and frefh air might probably be the means of contributing towards.

The hay confumed by this ftock was the produce of about fix ftatute acres.

The following information is from a refpectable farmer upon a large eftate about fix miles from Manchefter. He fays, that it will take two Lancafhire acres to fummer a milch

cow

cow about Chorton, and along the river Merfey, for eight or ten miles ; but that one Lancafhire acre in other places will produce not only fummer grafs, but alfo hay to keep a cow all the winter, if the fummer be moderately kind.   In the north of Lancafhire it will take three acres for each cow.

CHAPTER

CHAPTER IX.

Of GARDENS *and* ORCHARDS.

IN the neighbourhood of the large towns, there is a portion of land appropriated to Gardens.

Upon the banks of the Irwell, in the townſhip of Barton, about five miles from Mancheſter, there are ſixty-four ſtatute acres of land planted with apple-trees. The plants are upon borders of three feet wide, and ſeven yards diſtance from each in the rows, and from each other every way. The intervals in the rows, and between each apple-tree, are planted with pears, plums, cherries, and gooſeberries, which are intended to be removed as ſoon as they are found to incommode the apple-trees; and the borders are moreover dug, and cropped again with potatoes, beans, cabbages, &c. The intervals between each of theſe borders are under the following management: a part is appropriated to nurſery ground, for raiſing foreſt and fruit-trees; another large part is for meadow land, the graſs is mown for hay, and the eddiſh for ſoiling, and lets after the rate of 4*l.* 10*s.* per large acre. The plantation included in this acre ſome part ſown with grain. The plantation was begun about ten years ago, but was not completed till 1794, when the whole remaining was planted with crab-ſtock, to be in-grafted the enſuing ſpring. The trees look healthy in general, and if the kinds are well ſelected, and adapted to the nature of the ſoil, will moſt likely prove a beneficial concern in the iſſue, ſince Mancheſter and its environs will afford a ready market for an article much wanted, and but little cultivated.

It is generally believed, that there is not a town in the king-dom, London excepted, better provided with vegetables, roots, &c. than the town of Liverpool *.

* There are always ſome reaſons for diſtinguiſhed ſuperiority; and it has been ſaid, that the French neutrals (who were brought over from Ca-nada in the war of 1756, and who reſided ſome years in Liverpool) re-quired ſo many vegetables in their ſoups, &c. as to raiſe the market pr ce of theſe articles, which excited a ſpirit of growing greater quantities than had before been uſually raiſed. As a ſea port, the quantities of cabbage, and other vegetables, taken out for the uſe of, ſhipping; the quantities of dried herbs carried to Africa; and onions exported, may act as ſtimu latives.

Beſide

Besides the vegetables brought in by the milk-carts, and which really amount to a confiderable quantity; there is a certain farm in Kirkby, about eight miles north-eaft from Liverpool, the foil of a fmall part of which is a black loamy fand, and which produces great quantities of early, and ftrong, afparagus ; and another farm, a part of which is of the fame nature, at a place called Orrel, about four miles north-weft of Liverpool ; both which produce this plant with lefs at-tention, and lefs dung, than requifite in the rich vale of Kirk-dale, about two miles from Liverpool, where the greateft quan-tity of land in any place of this neighbourhood is appropriated folely to horticulture. In lands not favourable to the afparagus plant, might not this unfavourable difpofition be corrected by foil brought from lands more genial to its production, efpe-cially to grounds bordering upon the canals ?—Forty tons would be probably fufficient for a plantation for a moderate-fized family, and which when once matured continues for a number of years. This plant, in its wild ftate, is faid to grow upon the Bid-ftone Hills in Chefhire. The number of acres under horticul-ture in Kirkdale is about 28 of the large meafure *; and upon which are only employed about one male to each acre for the year, and one female to weed, and gather the crops of peas, fruits, &c. The mafters, it is true, are all workmen, and join with the labourers in their tafks ; by which is ef-fected, what otherwife would not have been accomplifhed, without a greater proportion of hands to the quantity of acres; and yet, fmall as this number at firft fight may appear, it is al-moft as wonderful how the mafter is enabled to pay his land-lord, his labourers, and his feedfman, their refpective claims, upon this portion of land, when the calculation is begun ; and 25l a year is allowed the man for his yearly labour; the half of that fum for the woman's; about 15l. more for rent and dung ; befides the expence of marketing, and the profits that fhould arife to the mafter for his attention, fkill, and fuper-intendance, and towards the maintenance of himfelf and family, with a fmall accumulating furplus, to fupport the infirmities of

---

* Eight yards to the rod, or to the pole or perch.

old

old age. In the amount of thefe feveral particulars enumerated, a fum of money will appear, that would have been fufficient to have purchafed the fee fimple of the fame lands, half a century ago.

The horticulture of this county is in many inftances fuperior to its agriculture. The mechanic is generally furnifhed with a fmall patch of ground adjoining his cottage; and from this little fpot is extracted not only health, but derived pleafure, and which may not a little contribute to fobriety; intemperance not unfrequently proceeding from want of recreation to fill up a vacant hour. This fmall fpace is devoted to nurturing his young feedlings, trimming his more matured plants, contemplating new varieties, in expectation of honours through the medium of gained premiums. Thus ftarting at intervals from his more toilfome labours, the mechanic finds his ftagnating fluids put in motion, and his lungs refrefhed with the fragrant breeze, whilft he has been thus raifing new flowers of the auricula, carnation, polyanthus, or pink, of the moft approved qualities in their feveral kinds, and which, after being raifed here, have been difperfed over the whole kingdom.

Not only flowers but fruit have been objects of their attention. The beft goofeberries now under cultivation had their origin in the county of Lancafter; and to promote this fpirit, meetings are annually appointed at different places, at which are public exhibitions of different kinds of flowers and fruits, and premiums adjudged. Thefe meetings are encouraged by mafter-tradefmen and gentlemen of the county, as tending to promote a fpirit which. may occafionally be diverted into a more important channel.

At thefe meetings, goofeberries have been produced which have weighed fingly 15 dwts. 10 grains; *e. g. Lomax's Victory**. *Woodward's Smith* * has weighed 17 dwts.; and the *Royal Sovereign**, grown by George Cooke of Afhton, near Prefton, at a meeting held 1794, weighed 17 dwts. 18 grains.

A fingle goofeberry-tree, the Manchefter rough red, in a garden belonging to Mr. J. Sykes, in Gateacre, in the year

---

* Names of goofeberries.

1792, yielded twenty-one quarts of fruit in their green state, when they sold at 3 *d.* per quart. The whole quantity weighed twenty eight pounds avoirdupois *. The space this tree occupied was three yards, and allowing an equal space to walk round, and supposing an acre of eight yards to the rod planted with the same kind of trees, and producing the same quantity of fruit, and sold at the same price, the produce would amount to £. 426. 16 s.

Requiring but little attention, the gooseberry has less paid to it than it deserves; and the fruit being rendered in such abundance, with so little trouble, makes it of trifling estimation. But since it may be improved in flavour, increased in quantity, and its duration prolonged, by being allowed a solitary corner in a wall, *e. g.* on each side the nectarine or peach whilst in their infancy, and they only occupy a small space; the gooseberry may be nailed down, trimmed, and trained as their companions; but removed as soon as ever they appear to incommode these ancient tenants of the walls; for the first cost of a gooseberry-tree is so trifling, that it is not worthy of notice.

These facts have been already proved by Daniel Daulby, Esq; of Birch House, near Liverpool, who for some years has had them planted against the walls, besides his other plantations of standards. Besides the advantages above noticed, the fruitage season may be advanced or prolonged according to the different aspects of the walls; and an increase of crop was thoroughly proved by this treatment in the year 1793, when there was a general failure throughout the kingdom, and gooseberries sold at the advanced price of 6 d. per quart. Those trees which had the advantage of walls were loaded as fully as in the most plentiful years.

* To ascertain the weight of this fruit in different states of its growth, the surveyor made the following experiments upon the Manchester red gooseberry.—1794, May 3, one ale quart weighed 18½ ounces troy.—July 25, again from the same tree 20 ounces.—July 15, 21½ ounces.—July 29, 22 ounces.—August 4, 21½ ounces.—*N. B.* He has to regret that he did not number the fruit.

Except

Except the orchard on the banks of the Irwell, in the township of Barton, containing about sixty-four statute acres, there
no orchards worthy notice.—There is no cyder made in the
county. The importation of apples from the cyder countries, and even from America, has of late been very considerable.

To cause fruit-trees to bear, particularly pears, cut a circle
through the bark round the principal branches.—This operation stops the growth of the wood, alters the system of vegetation, and gives the tree a tendency towards bearing fruit
instead of making wood.

The off-shoots of pear-trees should be taken off in August.

Lime dissolved in water, and made into a white wash, applied to the branches and stems of trees with a brush, effectually destroys moss *.

It is unfortunate that orchards are not more attended to in
this county, as cyder, with the assistance of honey, might be
made into a vinous liquor, as strong and as palatable as Madeira. The following is reckoned the best receipt for making it.

" Take new cyder from the press, mix it with honey till it
bears an egg, boil it gently for a quarter of an hour (but not
in an iron pot), take off the scum as it rises, let it cool, then
barrel it, without filling the vessel quite full ; bottle it off in
March. In six weeks afterwards it will be ripe for use, and
as strong as Madeira. The longer it is afterwards kept the
better."

Honey also renders hard crab cyder palatable. Colour and
flavour are easily added. Honey from the flower of the buck-
wheat may be made use of, if a dark hue is wanted.

There is every reason to believe, that currant, gooseberry,
and other home made wines, treated in the same way, would
equal what we are at such an expence in importing from foreign countries. The art of making it, with the assistance of
Father De San Martino's experiments on the fermentation
of vinous liquors (see Dr. Scandella's Addenda to the Chapter
on Manures) might soon be brought to such perfection, as to
make us independent on foreign nations for this important
article.

* In gardens where shallots are sown, to prevent the grub eating
them, they should be planted very ebb.

CHAPTER X.

WOODS AND PLANTATIONS.

THERE are no natural woods of any confequence to merit attention. The plantations are in general intended as
embellifhments for gentlemen's feats, cover for game, or fhelter from the blaft, rather than with a view of fupplying the
country with timber, and preventing importation.

Towards the coaft it is with great difficulty that wood
of any kind can be raifed: the tops of the trees, hedges,
and even the corn in the fields (in general) bend towards
the eaft, as if fhrinking from the weftern gale, brought over
the Atlantic ocean; yet, near the fhore at Formby Hall,
feveral acres of land have been planted with foreft and fruit-
trees, which are in fo flourifhing a ftate as to afford general
encouragement to the inhabitants of the fea-coaft, to fence
againft the wintry blaft, and to raife wholefome fruits for their
tables. The foreft trees were originally planted in holes
when very fmall, and were fheltered by fods from the winds
till they had taken firm root in the ground. A mixture of rich
foil and mofs was put with no fparing hand beneath their roots.
The Scotch fir, the fycamore, the platanus, and the afh, feem
moft congenial to the foil, which is of a fandy nature, and are
leaft injured by the inclemency of the climate. In the northern
part there are many acres of coppices cut down every fifteen
years, and burned into charcoal. Toward the central part of
the county there are fome good woods; the timber healthy:
there is alfo a confiderable quantity grown in hedge-rows;
but as fun-fhine is generally preferred to fhade—timber wood
feems on the decline. There are many excellent plantations
about gentlemen's feats and pleafure-grounds, well attended to,
fecured, and in a thriving ftate.

Mr. Leigh Phillips obferved, that the alder was of late
years become an article of great confequence, from the demand for its wood, which makes the beft poles whereon to hang
cotton yarn to dry, that wood acquiring a fine polifh by frequent ufe, nor does it fplinter by expofure to the weather, and
its bark alfo fells at nearly one penny per pound, as an article for
dye.

dye *. He added, that the alders planted on the fide of the Duke of Bridgewater's canal, upon the loofe grounds, for a certain diftance, by way of fecurity to the banks, had not only anfwered the original purpofe, but had proved a profitable plantation—the alder admitting of being cut down every fourth or fifth year. There are many acres of land, at prefent of little value, which, if planted with this wood, might probably turn to a good account.

The ofier willow is at prefent in fuch demand for hampers, &c. and there is fuch a fcarcity of that article, that more than twenty pounds a year have been made out of a fingle acre of land planted with it; and though very few acres are at prefent planted with them, there are fome thoufands proper for their growth, but the management of them feems not to be underftood at prefent.

On the fea-coaft there are fome acres of land planted with foreft-trees, which are flourifhing and ornamental to the country. They were originally placed in holes (with a mixture of fea-flutch and broken pieces of turf at their roots) four inches beneath the furface of the ground; and fods were raifed round them, to guard their tender fhoots from the wintry blaft. Its violence is leaft injurious to the fycamore, the afh, the alder, fir, and platanus.—This obfervation is communicated by the Reverend Mr. Formby, of Formby, who has fucceeded in raifing plantations fo near the fea, that it was hardly thought practicable till he effected it.

---

* In Sweden they make beautiful tables of the root of the alder.

CHAPTR

CHAPTER XI.

OF WASTE LANDS.

IN this county there are large tracts of waste lands, not less than one hundred and eight thousand five hundred acres, according to Mr. Yates's statement, who took the pains to calculate the number for this particular purpose —He makes the lands, under the denomination of moss, or fen lands, to be twenty-six thousand five hundred acres. Moors, marshes, and commons, to amount to eighty-two thousand acres. Why seek out distant countries to cultivate, whilst so much remains to be done at home?

At Lancaster there is an excellent salt marsh, adjoining the banks of the river Lune; and of which about 500 statute acres belong to eighty of the oldest freemen of the corporation of Lancaster, or their widows, and the trustees of this charity, the corporation. This marsh is pastured, and divided into what are termed *orl grasses*; that is, a privilege of turning one horse or two cows of any size to summer upon this common; so that a poney is reckoned equal to two oxen, however small the horse, or large the ox. The number of grasses or gates is equal to that of privileged burgesses, namely 80, and two more to the trustees of the charity, or 82 gates; and which, if let, are worth at present from £. 1. 10 s. to £. 1. 11 s. 6 d. per summer.—Seven years ago they would not let at twenty shillings a gate.

Now this marsh, if divided into fields of a proper size, is so fertile, that it would immediately be worth three pounds per acre; and, if improved, worth five pounds per acre per annum.

	£.	s.	d.
The present value is 82 summer grasses, at £.1. 11s. 6d.	129	3	0
And suppose the winter herbage worth　—　—	50	0	0
Total　—　—	179	3	0

But,

But, if inclofed, its annual
value would, at £. 3, per
acre per annum, be —

	£.	*s.*	*d.*
	1,500	0	0
Excefs	1,320	17	0

If improved, at £. 5. per acre,
would be — — 2,500 0 0

| Excefs | 2,320 | 17 | 0 |

Such ftatements cannot require any comment to recommend them to public attention, and that too in a neighbourhood of a town diftreffed for inclofed land; being bound up on one fide by this marfh, and on the other fide by a moor, which extends to the very borders of the town; a moor too, which manifefts itfelf capable of being rendered fertile land, as is evident from fmall inclofures under cultivation, which the induftry of fome cottager has improved from the wafte.

In the neighbourhood of Prefton lies Prefton Moor, about 500 acres of good land, and abounding with excellent marle, but which at prefent lies under water, which might be eafily removed. Fullwood Moor, too, in the fame neighbourhood, about 1000 acres, and Caddeley Moor, which belongs to the crown, with many more which might be enumerated, and which remain in a ftate that difgraces the county.

Many of thefe lands are incapable of tillage—fome confift of mountainous tracts, craggy, fteep, and barren: thefe are employed for fheep walks, though not the moft fertile: others confift of low fwamps, overcharged with ftagnant water: from which a fufficient fall has not yet been difcovered for draining them. Many of the waftes are covered with underwood, and others have been planted with various kinds of foreft trees. Sir Harry Hoghton propofes to plant Withnell Moor, a tract of about eight hundred acres, with fuch trees as upon trial fhall be found to agree with the foil. Several parts are allotted out in what are termed dales, for the purpofe of paring the furface for fuel—a pernicious practice, which injures the land, and affords but a very indifferent fire.

There are many thoufand acres capable of being cultivated, and made into either arable, pafture, or meadow land, of a very good quality, provided thofe waftes were inclofed, di-

x   vided,

vided, and improved; and, to effect this, there is neither want of inclination nor spirit amongst the inhabitants. But there is a want of a general inclosure bill to facilitate that troublesome business, and render it more expeditious and less expensive.

A great improvement has been suggested by Mr. Wilkinson, of Castle-Head, of embanking upon the sands, and gaining thereby 30,000 acres. This great attempt has been already noticed in the Annals of Agriculture; but these patriotic and public attentions are at present defeated, by a difference of opinion amongst individuals, claims of the lords of the manors, &c.

Mr. Wilkinson also, by turning the course of some brooks, has recovered lands from the sea by which the flux of the tide, in the space of about eight years, has raised the lands near six feet; so that, after the water is kept in narrower bounds, by the opening of a new channel, the tide alone does the work.

OBSERVATIONS ON THE EMBANKMENT OF LANCASTER SANDS.

" IT is a fact, consonant to reason, and proved by experience, that when the course of a river where it enters the sea, or rather tide mark, is turned another way into the ocean, the former channel, and adjacent sand, is, from the perpetual influx of sand, mud, &c. brought and left there by the tide, raised gradually, till, in the course of a few years, it becomes out of the reach of, at least, ordinary. tides; because the fresh water ceases to prevent the accumulating of these materials, which it formerly did, by constantly removing them to the sea.

" If that is the case, there must exist a possibility of recovering from the dominion of Neptune that extensive tract called Lancaster and Milthrop Sands; as also, part of the Ulverstone, and Dudden or Millam Sands, by a diversion of the rivers.

" The first question naturally arising in the enquiry is, Whether an effectual removal of the rivers is practicable? and, secondly, Whether, in that case, the probable expence would not

over-

overbalance the advantages that might be expected to arise therefrom?

"In regard *to* the firft: an ingenious and refpectable gentleman in that neighbourhood, Mr. John Jenkinfon of Yealand, had, for many years back, given the fubject much attention, and minutely explored the track propofed for the new channel of the Kent and other rivers running through the Lancafter and Milthrop fands, as pointed out in the plan. Some years fince he communicated his ideas on the matter to Mr. Wilkinfon of Caftlehead, a gentleman of fortune, patriotifm, and univerfal knowledge. The fcheme attracted the notice of Mr. Wilkinfon; he examined the ground, and was immediately ftruck with the notion that it might be carried into execution without much difficulty. A fubfcription was propofed, in which Mr. Wilkinfon offered to lead off with 50,000 *l.* if the neighbouring gentlemen would make up the reft (having previoufly eftimated the whole expence at 150,000*l.*) or, if they would begin with any fum, he would produce the remainder, it being underftood that each fhould receive of the profits in proportion to his fubfcription. The project being thus apparently pretty forward, a perfon was appointed to take the levels, &c. which he did; and his plans are now in the poffeffion of Mr. Jenkinfon, who alfo himfelf made an actual furvey of Lancafter and Milthrop fands, from whofe plan I copied part of mine.

"Notwithftanding thefe preparations, the projectors unfortunately met with fuch oppofition from the proprietors of fome trifling fifheries, who were neverthelefs offered an indemnification for the lofs they might fuftain; and certain lords of manors, who, though they refufed to contribute any thing towards recovering the fands, were yet unwilling to relinquifh any part of their claims to the ground when improved—that the matter was dropped at that time.

"The principal river to be taken off Lancafter and Milthrop fands is the Kent. I examined with attention the ground propofed for the new channel, as marked in the plan, and found it remarkably adapted for the purpofe. The whole length, where it runs inland, is a range of low mofly or foft land, except a

N   fmall

small tract of rocky or gravelly ground, the highest part of which is not more than 10 feet 5 inches above level; and I believe the average height of the whole cut would not exceed 3 feet 5 inches above level. In short, I do not entertain a doubt of the practicability of diverting the course of the rivers, and taking them into the Loyne, below Lancaster. The fall in that course is small, yet sufficient for the current of the water. Neither do I find a difficulty in believing that the ultimate consequence would be the gaining a very large tract of sand, which would become the finest land. This method of recovering ground from the sea is now, where it is practicable, universally allowed to be a much surer, and often less expensive, means than that of wholly depending on embanking on the sand with any materials whatever.

" Whether it would be best to follow exactly the plan I have prescribed, in diverting the rivers, is the province of an experienced engineer to determine. Equal knowledge and abilities are required to make a tolerably exact estimate of the expence in the execution of such a design. I shall, however, from all the knowledge I could possibly acquire of the business, endeavour to make out an estimate, which may, at least, convey a general idea of the scheme; but which, my inexperience in these matters bids me add, must not be too implicitly relied on in particular.

" Mr. Wilkinson, as observed before, calculated the whole expence at £. 150,000; but in the opinion of many well-informed gentlemen 50 or perhaps £. 60,000 less might do. Various plans have been proposed by different people; but it would seem best, in my opinion, to commence the work a little below Dallam Tower (as shewn in the plan) by throwing a bank of stone, or stone and brushwood, across the channel there: plenty of these materials being at hand, on a common. The bank would serve for a road, and a bridge at the S. E. end would admit the fresh water. The sand here is near thirteen feet deep, which it would be necessary for the stones to bottom; that would require little or no labour, more than tumbling in; as the weight of the stones and washing of the tide would soon bring them to the channel. This bank would

be

be about 880 yards long, and fhould I believe be 7 yards high, 10 yards at the bafe, and 6 yards at top, and would confequently contain 49,280 cubic yards, which, allowing each yard to coft one fhilling, would amount to £.2,464. The bridge I fhould ftate at £.1,000. The whole length of the cut from hence to the Loyne is about 21,340 yards: to contain the greateft land floods it fhould not, I prefume, be lefs than 34 yards wide, and the average depth 4 yards; the number of yards, upon that pofition, to be excavated, would, therefore, be 2,902,240, which at .4½*d. per* yard would coft £.54,417. Where rocks or high ground upon the coaft renders it neceffary to keep within the tide-mark, the earth to be taken out will form a bank on the fea-fide of the cut. A number of bridges might be neceffary to erect; however, till the profits of the land to be recovered fhould enable the proprietors to build them of ftone throughout, I fhould propofe temporary bridges of wood, except one, for the principal road; the expence of which we fhall call £.1,000, and that of the wooden ones £.3,600.

"The next thing to be confidered is the diverfion of Lindlepool, which might either be brought into the Kent, as fhewn in the plan, or taken the contrary way into Cartmel fands. In either cafe, as it is an inconfiderable rivulet, and the ground generally very low and foft, I fhall not ftate the expence at more than £.5,300, including the neceffary bridges.

" Afterwards, when the fea had nearly embanked itfelf, it might be found convenient to raife fand banks a few feet high, in order to keep off high fpring tides: the expence of which, added to that of purchafing ground for the new channels of the rivers, I fhall ftate at £.13,000.

" Thefe fands are the principal objects of attention, but fhould their recovery be effected, it would be found very convenient, as well as practicable, to ufe fimilar means in obtaining part of the Ulverftone fands. A bank might be thrown over the channel, as marked in the plan, with a bridge at the end of it, the frefh water then confined to the fhore till it entered Ulverftone mofs, through which an eafy cut would bring it to the fands again either at Plumpton Hall, or at the mouth of the new

N 2

canal, where there is plenty of rock at hand to fecure it. At the latter place it might be of fervice to the fhipping, by opening the channel.

" By that operation, about 1,600 acres would be gained. Every expence attending which I eftimate at £. 20,000.

" The acquifition of at leaft 4,600 acres may alfo be effected by the fame means upon the Dudden or Millam fands. A long ftrip of marfh land extending along each fide renders the tafk of diverting the rivers, comparatively, an eafy one. The Dudden might be conveyed along the north fide, and fixed, at its entrance into the fand, with limeftone rock : while the rivulet called Kirby-pool might with little obftruction be taken down the other fide, if we except the intervention of a little rifing rocky ground extending about an hundred yards. That, however, is no object in a work of fuch magnitude. The whole expence of this undertaking I am perfuaded would not exceed £. 26,000.

Let us now collect the feveral fums eftimated :

Expence of the bank below Dallam Tower - £.	2,463
Ditto of the bridge at the end thereof - -	1,000
Ditto of the cut from thence to the Loyne -	54,417
Ditto of the bridges over the cut - -	4,600
Ditto of fand banks, and purchafing ground -	13,000
Ditto of diverting Lindlepool - - -	5,300
Ditto of gaining part of Ulverftone fands -	20,000
Ditto of gaining part of Dudden fands -	26,000

Intereft of money funk, till the land to be gained becomes profitable; falaries of engineers, &c. with contingent expences, I fhall call   -   -   73,219

Total expences   -   -   -   £.200,000

The land that might reafonably be expected to be gained upon the Lancafter, &c. fands, is - Acres 32,510
Ditto upon the Dudden fands   -   -   -   4,600
Ditto upon the Ulverftone fands   -   -   1,600

Total number of acres   38,710

" We

" We are now to confider what benefits would accrue from the execution of the above projects.

" In the firft place, a regular connection would take place between Lancafter and Whitehaven, by a poft road, which would doubtlefs be laid out between thofe places; by which not only thefe commercial towns, but all the intervening country would be much benefited. Whereas at prefent, a perfon travelling between Lancafter and Ulverftone, Ravenglafs, Whitehaven, &c. muft either take a very circuitous rout through a wild mountainous country, or wait a precarious, dangerous paffage over the fands. A reflection on the number of unfortunate people, who are annually loft in croffing thefe deceitful fands, touches the nerve of humanity. That dreadful circumftance would be remedied by banifhing the tide. But although the philanthrophic mind may confider thefe matters as great grievances, others may look upon them as provincial evils only, and the effects of their removal equally confined. Another advantage that would take place would be more univerfally felt. Here are tracts of fand containing 38,710 acres, which at prefent, inftead of being beneficial to the community, are a general nuifance. If this land could be recovered by laying out the fum of £. 200,000, it would be a purchafe of £. 5. 3 s. 3 ½ d. per acre of land, which, I prefume, by the time all the money was paid, would be worth £.40 per acre, confequently a clear gain of £. 1,348,400.

" This would not be like a transfer of property, where one party lofes what the other acquires. It would be a property really gained, the produce of which (whoever were the immediate poffeffors) would expand itfelf, on every fide, to a great diftance; and by caufing an increafe of provifions, muft proportionably affect the price; whereby thoufands of poor families would find an additional morfel to their daily pittance, exclufive of the employment it would afford them in the execution.

" In hopes a little farther fuggeftion may not be offenfive, I fhall obferve, that, fhould the project be attempted, it would be prudent, or rather neceffary, after it is afcertained in whom the prefent property of the fands abides, with the affiftance of parliament, to require the proprietors either to contribute their

quota

quota towards the expence of obtaining the fame, or for ever to forfeit their right thereto, which fhould be transferred to the firft who offered to make good the fubfcription.

" As Mr. Jenkinfon, mentioned before, is perfectly acquainted with the place, and nature of the fcheme, he would be a very proper perfon to apply to by any gentleman, wifhing to have a further knowledge of the fubject, in any particular."

M O S S E S.

IN the parifh of Eccles, is a large tract of mofs land called Chat Mofs, lying between the townfhip of Worfley and the navigable river Irwell, containing fome thoufand acres; and on the fouth fide the river is another piece of land called Trafford Mofs, which adjoins to the park of John Trafford, Efq; and contains about 500 ftatute acres.

Thefe lands, which have hitherto been totally uncultivated and of no ufe whatever, except that of fupplying the neighbourhood with peat or turf for fuel, are advantageoufly fituated for improvement. The country round is populous: Chat Mofs approaches within fix miles, and Trafford Mofs within three miles of Manchefter. The Duke of Bridgewater's canal divides Trafford Mofs, and terminates at fome diftance in Chat Mofs. The lands lie upwards of thirty feet above the bed of the river; and materials for improving them, when drained, are found in many parts of the neighbourhood.

The nature of mofs lands is too well known to require any defcription—they have probably originated from pools of water fed by adjacent fprings or rain, which from the peculiar conformation of the ftrata below, have not been able freely to trace a fubterraneous paffage, and have become ftagnant. In courfe of time, thefe pools admit of vegetation of various kinds, which having annually fubfided, afford a proper fubftance for the nutriment of fuch other plants as are ufually found in thefe

fituations,

fituations, which, befides the various fpecies of mofs, the growth of fome of which is aftonifhingly rapid, are the *erica vulgaris*, the *ornithogalum luteum*, and the different fpecies of *eriophorum* or cotton grafs.—As thefe plants decay and depofit their fubftances, a confiderable addition is yearly made to the mofs, in cutting a fection of which it is not difficult to perceive, and to divide from each other, the vegetation of each year, which appear in lamina growing more indiftinct, hard, and cohefive, according to the depth of the mofs. The plants before-mentioned, and particularly the mofles, feem to find their proper nutriment in their own ruins, and grow more luxuriant as the fubftance of the mofs increafes ; at length the whole takes the appearance of a large fungus or homogene vegetable : continuing to increafe, it at length rifes greatly above the level of the adjacent lands, till the weight of the furface becoming too great to be fupported by the fpongy fub-ftance below, it begins to overflow its banks, and cover the adjoining grounds, as happened of late years at Solway Mofs, and was formerly the cafe at Chat Mofs, a great portion of which detached itfelf into the Irwell ; and, if we may be-lieve our ancient chroniclers, was carried by the Merfey into the Irifh fea.

In the year 1793, Mr. Wakefield, and Mr. Rofcoe of Li-verpool, undertook the improvement of thefe lands, and a contract was entered into with the proprietor, Mr. Trafford, for a leafe of them for a term of years under a yearly rent. An act of parliament was obtained, enabling the proprietor to leafe the fame ; and the improvement of Trafford Mofs was immediately begun by interfecting it with drains at fix yards diftance, which opening into wider drains at one hundred yards diftance, convey the water arifing from the mofs into the river Irwell.

In cutting thefe drains, one precaution has been found of the utmoft importance. If the drain be cut to its intended depth at one operation, it will be impoffible to prevent the fides from falling in, and no labour can afterwards effectually repair the damage.

It is highly neceffary, therefore, to attend to the nature and
confiftence

confiftence of the mofs, and not to cut deeper at one time, than will fuffer the fides to remain perfectly firm. The method adopted on Trafford Mofs is, to open the drain at the firft cut only about one foot deep, which is thus left to drain, and at proper intervals is cut again till it is three feet deeper, and about eighteen inches wide; by thefe means the fides of the drain become not only hard and finer, but are perhaps of all other materials the moft durable; being unaffected either by moifture, froft, or fun. When the drain, thus cut, has remained fo long as to have become tolerably dry at the bottom, a narrow drain is opened in the middle of it with a fpade, about five inches wide and eighteen inches deep, which thus leaves a fhoulder of about fix inches on each fide, intended for the fod or turf, with which the narrow drain is covered, to reft upon. The narrow or fplit drain is then carefully cleaned, and covered with the firft fod cut from the drain, the furface or fwarth being turned downwards; and the whole is then covered up ready for cultivation. A confiderable part of Trafford Mofs is thus drained, and the reft is interfected with drains at fix yards diftance, a great part of which will be covered in the prefent year. In confequence of thefe operations, the mofs has funk confiderably, and acquired a great degree of folidity.

This operation being completed, the furface of the mofs is to be levelled, and the fod turned under, which may be done either by the pufh-plow, or the fpade, both of which methods have been tried at Trafford; but the latter, though a more expenfive operation, is thought to be preferable, as the tough fod is thus effectually covered, and a furface produced, which admits more readily the operation of the air, and more eafily mingles with the materials employed in the propofed improvement.

The materials which have hitherto been chiefly tried, are fand and marl, both of which are found at the fouthern extremity of Trafford Mofs, the latter of an excellent quality. Thefe have been ufed together (laying on the fand firft), and feparately, and it is expected the effect of each will, in fome degree, be afcertained in the courfe of the pre-

†   fent

fent year. The land not being fufficiently hard, in the firft ftages of improvement, to allow the materials to be conveyed in carts, the undertakers have availed themfelves of a road made of iron, caft in bars of fix feet long, and jointed together by dove-tailed fteps, refting upon wood fleepers. Upon this road one horfe will with eafe take feven waggons of marl or fand, of fix hundred weight each. The extremity of the road, where it diverges on each fide from the principal road, is daily changed; and a fingle perfon will, with eafe, take up, remove, and lay down two hundred yards of it in a day. A fpace of fixteen yards wide, or eight yards on each fide the road, is then covered with the materials employed, beginning with the furtheft extremity of the road, and as the work proceeds from thence towards the main road, a perfon is employed in taking up the moveable road, which is of no further ufe, and removing it to the diftance of fixteen yards, by which means it is in readinefs to begin upon as foon as the marling, or the former road is completed. The horfes have relays at proper intervals, and the marl is thus conveyed to the furtheft part of the mofs.

Of the effect of thefe operations, it is yet premature to fpeak. About ten ftatute acres of potatoes were laft year planted in the mofs, manured with the common town foil of Manchefter, and produced a very good crop.

The fame land has fince had a cover of marl, and is fown with barley; about twenty acres of the marled land have been fown this fpring with vetches, and the other parts of the mofs in cultivation are principally cropped with potatoes and oats.

The following engravings will explain the nature of the operations above defcribed.

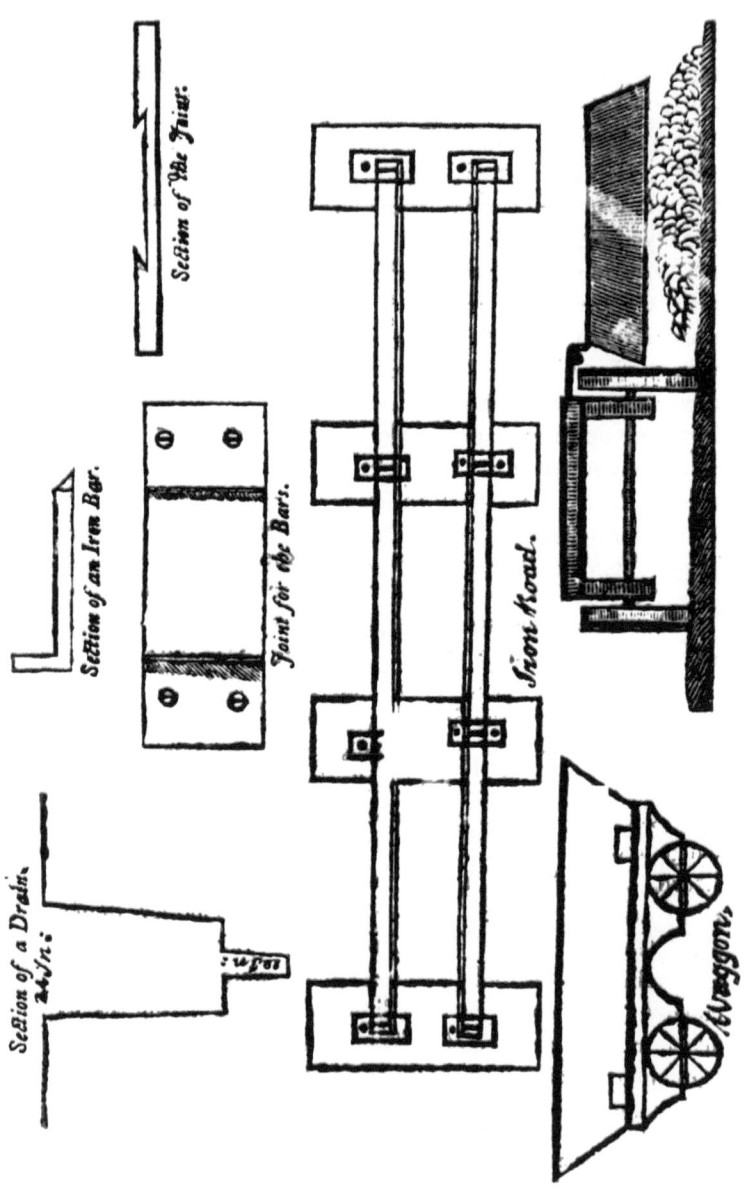

Section of the Frize.

Section of an Iron Bar.

Joint for the Bars.

Section of a Drain.

Iron Road.

Waggon.

RAINFORD

RAINFORD MOSS.

MR. JOHN CHORLEY of Prefcot, having taken a part of Rainford Mofs, belonging to the Earl of Derby, upon a leafe of three lives, and at a rent of eight fhillings *per* acre, *per annum*, befides a fmall fine, began to improve the fame in the year 1780. The land is a poor barren mofs, not of the leaft value in its natural ftate, being fo fpongy and full of water, as not to admit the foot of cattle upon its furface. After draining, by open drains, three feet wide at top, to the depth of two feet, and afterwards one foot deeper, and only nine inches broad at the bottom, the interval between each drain eight yards, the expence of cutting which was three-pence for every eight yards, he began with pareing and burning, with crops of oats, barley, and clover; till being convinced of its *deftruΔive effeΔs* (to make ufe of his own expreffion) not only upon his own, but from the experience of others in the neighbourhood *, he totally abandoned that practice in 1787, and has adopted (amongft others which he has re-gularly regiftered in a book he keeps for that purpofe) the following courfe, copied from his memorandums. Pota-toes with dung, for the firft time, produce about four hundred bufhels *per* large acre of eight yards; next year potatoes again without dung—produce about three hundred bufhels. He is this year (1795) trying potatoes for a third time, without dung, and feems to fpeak with confidence of fuccefs. To return, in 1789, upon the lot under notice, he fowed Tarta-rian oats, the produce handfome—but Mr. Chorley thinks mofs lands in general not proper for grain, being more favourable to the production of grafs, which comes fpon-taneoufly, if encouraged by a little dung—and he intends to difcontinue the practice of fowing grain; he fows his clover without any grain. His practice at prefent is to fow the clover immediately after the potatoes are taken up, if early in the

---

* The land he has improved without paring and burning, certainly has a fuperior appearance to that of his neighbours, who continue the practice; but that may be owing to their exhaufting the land by too many crops of corn after they have pared and burnt.

feafon.

feafon. Along with the oats was fown clover, and in 1790 two handfome crops were taken; 1791 mown, afterwards marled, about one rod of fixty-four cubic yards, laid upon an acre; 1792, 1793, and 1794 mown. The eddifh was not eaten off, but harrowed and raked away in the fpring, and ufed as litter for horfes.

His manner of planting potatoes (which are fet always the firft year when the ground is broke up) is as follows:—The mofs taken from the drain is put into the middle of the butt or ridge, and dug under, in order to raife it higher than the fides. The fpit is about twelve inches in depth, the expence 7 d. per rod.—After being dug and expofed to the air, the fur-face is broken with a fpade (expence 2 d. per rod) the butts or plots, of eight yards broad, are divided acrofs into ridges of two feet, acrofs which are planted three, but fometimes only two potatoe fets, upon which, or over the fets, is laid the dung; and over the whole is thrown the mofs, a foot on each fide being referved for covering when firft planted, and another for covering when the ftems appear above the furface—the whole breadth of each being four feet.

The

The whole will be best understood from the following sketch :

PLOT or FIELD of 8 Yards.

One foot	- - - - - - - - - - - - - - ·	Gutter or interstice between, from which the covering
Two feet	Potatoe sets o o o o o o	is taken.
One foot	- - - - - - - - - - - - - - -	Another interstice.

```
          o   o   o   o   o
          o   o   o   o   o
          o   o   o   o   o
    - - - - - - - - - - - - - -
    - - - - - - - - - - - - - -
              o   o   o
              o   o   o          Two potatoe sets only.
    - - - - - - - - - - - - - -
    - - - - - - - - - - - - - -
```

The expence of draining, digging, dung, planting sets, &c. for an acre of potatoes, he estimates at £.50 *per* acre, but thinks he is repaid the whole in the course of three crops.

Mr. Chorley has about thirty large acres under cultivation, about ten more ditched out, and about twenty acres under potatoes.—He prefers good horse or cow-dung to marl, which he thinks should not be laid on till after two or three crops—after it has lain some time under grass, it begins to run wild, and requires turning over again.—By a change of his potatoe sets from this moss, to his old inclosed lands, Mr. Chorley preserves his crops from the *curl*.—His sets are become famous on that account, and readily purchased for the purpose of planting by his neighbours.

It is with regret we add, that the curl is a general com-

---

* The dung brought from Liverpool costs him 10 *s.* 8 *d. per* ton when laid down upon the moss.

plaint this year (1795); that there is greater appearance of this difeafe amongft the potatoe crops than have been obferved for fome years paft.—Recourfe muft at laft be had to the feed, for renewal;—bulbous roots, it has been found by experience, decay after a certain number of years—" Ranunculus in " twenty-five, anemone in fifteen, and hyacinths in twenty-fix " years *." After which period, no art and pains can pre-ferve them, though a change of foil in the mean time is ufe-ful. It is proper however to remark, that the curl may be prevented from fpreading, by taking away any plants the inftant they feem to be affected with that difeafe. This important dif-covery ought to be known as generally as pofiible.—The quef-tion was put to Mr. Chorley; and he anfwered, that his crops appeared clear, nor did the furveyor obferve any infection.

He propofes to continue planting potatoes another fortnight from this date (15th June) and has at leaft thirty perfons em-ployed, men, women, girls, and boys, at this work.

He has built nine cottages, which he has named *Cheapfide*, as habitations for the labourers.he employs;—he only charges them with 20s. *per annum* of rent.

Profit from improving Wafte Lands.

Bootle Marfh, in the neighbourhood of Liverpool, was let before being improved at ten fhillings per acre, and is now worth about £. 3.—Trafford Mofs was formerly not worth one fhilling per acre; but fuch of it as has been drained, is now reckoned worth about £. 3 per acre per annum.—Bol-ton Moor, after an act of inclofure in 1793, was divided into lots, and only 170 ftatute acres was difpofed of for the immenfe ground rent of £. 2,600 per annum. But it was in fome meafure intended for building. Some of the lands in this moor have fince been cultivated. One inclofure was covered about two inches with foil; fown with vetches, without ploughing. An excellent crop. White clover fown amongft the vetches. The prefent year (1795) a very good pafture. In 1794, in-clofures of 12 ftatute acres produced 600 bufhels of oats, Win-

* See Madox's Florift's Directory, p. 91.

chefter,

chefter, which fold at 3 *s.* 6 *d.* per bufhel. Cultivation, one furrow, manure, a compoft of lime and earth. 4,000 bufhels of potatoes grown upon this moor 1794. Before inclofure, the furface of infignificant value for pafture. Produce only coarfe bent grafs. Under ftratum, clay, from which bricks were made.—Dean Moor lies contiguous, about the fame fize, and nearly as valuable; and near Bolton alfo there are other moors, capable of being improved at no very confiderable expence, and rendered worth four pounds per acre.—Kearfley Moor is very extenfive, fome bad, and fome exceeding good land; moft of it capable of cultivation, and contiguous to marle and lime. At prefent, being overftocked, the cattle ftarved, and of little advantage to the owners.—An act has been obtained for inclofing Edgworth Moor the prefent feffions. But the vexatious trouble attending this work, operates as powerfully as the expences of obtaining the act. If inclofed and improved, it would add much to the produce of this county.

Whitworth Moor alfo, near Rochdale, a very large tract, is capable of improvement, and of being rendered good land.

Many of the moors, if only inclofed (which, in their prefent ftate, are of little confequence) would immediately become of very confiderable value.

CHAPTER

CHAPTER XII.

IMPROVEMENTS.

SECT. I.—*Draining.*

THERE has been much draining done in many parts of the county; but there remains much still to be done: but the spirit is gone forth, and the good effects are evident, so much so, that in many instances that have been mentioned, the land has been so far improved, as to repay the costs by the superior crops which followed this improvement, even the very first year, after the work was executed.

All draining is trifling, in comparison of the practice of Mr. Elkington, of Prince Thorp, near Coventry, who is now employed in many parts of the kingdom with surprising success.

The mosses in general might be effectually drained, and at a small expence, were the springs that feed them cut off and carried away from the high lands before they reach the mosses. Mr. Elkington has improved several as above, and rendered the lands of great value.

Were Mr. Elkington's principles of draining made public, this county would in particular be benefited by his discovery.

" The cheapest and most effectual method of improving " moss lands, as Mr. Taylor justly observes, is that practised " by Mr. Elkington, who discovers and carries off the springs, " that cause the bogs."

His system is so simple and so rational, that it strikes with immediate conviction. As the Board is already in possession

of

of the principles of his mode of draining, it is unneceſſary to dwell longer upon the ſubjeċt. One example of its importance, however, it may not be improper to give, though on a ſmall ſcale. A ſingle drain in a field of four acres of the large meaſure, was calculated to coſt four pounds. The advantages to be derived might be reaſonably eſtimated at not leſs than eight pounds per annum upon that field alone; but its beneficial effeċts probably extend beyond the limits of one ſingle field; to what extent, further experience will prove. The ſource of a wide ſpreading evil is thus, with one ſtroke, diverted into another channel, and its bad effeċts totally cut off.

J. Wilkinſon, Eſq. on the borders of the county, has drained to the amount of 1,000 acres of fen lands; Warton Moſs has alſo been drained. Trafford, and a large part of Chat Moſs *, are taken by Mr. Wakefield and Mr. Roſcoe, on a long leaſe, with intention to drain. Near one hundred acres are already cut upon Trafford Moſs, upon which Mr. Wilkinſon's plan is purſued, of making uſe of the materials upon the ſpot; cutting through the moſs at different intervals of time; by which is given opportunity for the water to eſcape, the ground to acquire more firmneſs, the walls to grow harder; and as the ground would otherwiſe cloſe, at a diſtance from the bottom, a large ſhoulder is left, whereupon a lintel is to reſt, cut from ſome ſolid turf, about 18 inches in length, and 9 inches ſquare, and which, being expoſed to the ſun and air, contraċts its dimenſions to nearly one half, acquires firmneſs, hardneſs, and ability to ſupport the matter with which the ſurface of the drain is covered.

The fens or moſs lands thus drained have acquired ſolidity, and become fertile meadow, and corn lands; and, in conſe-

---

* " Chartley-More braſt up within a mile of Morley-hall, and deſtroyed much grounde with moſſe thereabouts, and deſtraid much freſch water fiſch thereabouts, firſt corrupting with ſtinking water Glaſebrooke, and ſo Glaſebrooke carried ſtinking water and moſſe into Murſey water, and Murſey corrupted, carried the rowling moſſe part to the ſhores of North Wales, part to the Iſle of Mann and ſum into Ireland. In the very topp of Chartley-More, where the moſſe was higeſt and brak, is now a plane valley as was in tymes paſte, and a rille runneth in hit, and peaces of ſinaul trees be found in the bottom."

quence of the drainage, have funk fome feet lower *. Warton mofs, and Mr. Wilkinfon's, are become very rich meadow and pafture land.

The only effectual means of fpeedily forwarding irrigation throughout the kingdom, to the utmoft extent, would be to eftablifh a company of able practitioners in that line, to whom individuals could apply for advice, to direct the works in the beft and moft effectual manner, or who would undertake to compleat the whole for certain fums *per* acre, according to fituation, &c.

If the above affiftance could be eafily obtained, there is no doubt but that thoufands of acres would be turned to that moft valuable mode of management, in the courfe of a very few years.

The fame may be faid of draining, and were Mr. Elkington's (or any, if poffible, other fuperior mode) made public, and a company able to direct, formed, there would not appear in this county fo many thoufand acres of morafs, within a very fhort period.

There is a variety of drains befides the above; a piece of peat, the ufual fhape and dimenfions of the common turf, has been made ufe of, after piercing the turf with a kind of punch when wet, by which a hole is left about three inches fquare, a little arched at the top in this form ⌒, and after being hardened in the air, the two pieces of turf are placed fide by fide. For this the Agricultural Society at Manchefter rewarded the inventor with a premium.

Common brick, with thin flates at the bottom of the drains, have been frequently ufed. A double brick, with a hollow through the middle, is an article cheap, foon made, durable, and fufficient for the purpofe. Broken ftones have been frequently ufed, laid loofe and open, the drain firft cut in this

form ⎣ ... ⎦, and filled up as far as the dotted line. But the

* Mr. Wilkinfon's mofs is, in fome parts, fuppofed to be funk fix feet lower :—before the drainage, the windows of the third ftory of Mr. Wilkinfon's houfe juft appeared from a certain point; but from that place, at prefent, the windows on the firft floor are plainly feen.
    Since writing the above, Mr. Wakefield obferves, that an actual meafurement has been made, and the fall of the mofs is about four feet and a half.

cheapeft

cheapeft are the fod drains, made by T. B. Bayley, Efq. of Hope near Manchefter. The implements and manner are particularly defcribed in Dr. Hunter's Georgical Effays. I viewed the drains, which have already ftood twenty years. The entrances have generally a fence of brick, or ftones, to fecure them from the feet of cattle. This work is performed at fixpence per rod: men were employed in cutting new drains when this well-managed eftate was furveyed.

More attention fhould be paid to draining marle-pits than is generally practifed; the ftagnant water frequently overflows, and ftarves a large fpace of land, till its effects are deftroyed by fome ditch, &c. which cuts off the nuifance by carrying the water off *, but the draining of the pits not only removes this evil, but is the means of gaining a confiderable fpace of ground.

A good practice, by S. Fazakerly, Efq. fhould be noticed. When fall fufficient into the main drain, to take off the water from fome particular fpots, is not afforded, he finks a kind of well where the fpring arifes, the fide of which he fecures by ftones or brick, and thus collects the ftagnant water into one point, and by this means he can get rid of it. Mr. Bayley of Hope mentioned an improvement upon this mode, namely, an auger-hole has been found effective if properly applied.

Mr. Ecclefton has applied his miner, this prefent year, for the firft time, with apparent fuccefs. The furveyor walked over a field where the miner had been drawn through certain intervals, only once; the run of water was not trifling, and the ground feemed firm.—The expence of this operation is very inconfiderable.

Obfervations on the BRICK TAX, *by* T. B. BAYLEY, *Efq.*

" Very important and *extenfive* fchemes of draining *moffes,* &c. in this county are projected, and depend on taking off this tax; and I have frequent applications on the fubject, as the feafon for making bricks approaches. The prefent feafon muft furely convince every man who has eyes to

* J. J. Atherton, Efq. has done much in this way.

fee,

ſte, that a ſpirited agriculture muſt be our final and beſt re‑
ſource. Perhaps the moſt ſimple mode would be to allow a
drawback of the duty for bricks uſed in draining; though I
have a great objection to the principle of drawbacks, as temp‑
tations to fraud—*effected* by *perjury.*

 " From the quantities of rain which fall in Lancaſhire, and
the nature of our ſoil in general, draining is, of neceſſity, the
firſt requiſite ſtep to improve our lands. Moſt parts of the
county have not any *ſtone,* and the tax on *bricks* has operated as
a *total prohibition* of *their* uſe in draining. This circumſtance
has been of the greateſt poſſible diſadvantage to our agricul‑
ture, and was communicated by our county members and other
gentlemen to adminiſtration laſt year, when the new duty was
laid on bricks. The repreſentation was kindly received, and
attended to — The impolicy of obſtructing the *means* of na‑
tional improvement, eſpecially of its agriculture, was ſeen
and acknowledged by the ſecond ſection of the 34th Geo III.
chap. 15. But the great ſize and preſcribed ſhape of the *tile*
or brick for draining, effectually prohibits its uſe, and takes
away the indulgence meant for us.

 " A common brick of the uſual ſize and ſhape is, on every
account, beſt adapted for *draining,* as it forms the *bottom,* the
walls, and the *covering* of the drains * , and I really think the
revenue would not be injured, if the legiſlature was generally
to exempt bricks made for the *expreſs purpoſe of draining* from
the tax. To prevent frauds and abuſes, perſons might be ſtill
obliged to enter their bricks at the exciſe office, and to pay a
ſmall duty of three pence per thouſand to defray the expence of
the officer's attendance, and be ſubject to a *very heavy penalty*
for applying thoſe bricks to any other uſe or purpoſe than that
of draining; they might be further required to certify to the
exciſe officer the time, place, and manner in which theſe bricks
are uſed.

 " The uſe of ſo bulky a material as brick cannot be eaſily
ſmuggled; numbers muſt be privy to it, and the fear of detec‑
tion, and of a heavy penalty on the owner and *workmen,* would,

---

 * *Bricks* are chiefly uſed in main drains. or ſoughs laid at conſiderable
depths, and have ſtrength to bear a weight of earth which tiles have not.

I am

I am perfuaded, *totally* prevent all illicit attempts to defraud the revenue.

" Should not this fimple expedient be adopted, perhaps fome irregularity in the fides, ends, and alfo the furfaces of the draining bricks may be devifed, which might not at all unfit them to form the *bottoms, walls*, or *coverings* of a drain, and yet render their ufe in building difficult or impracticable, as A

" This circumftance is worthy the IMMEDIATE attention of the Board of Agriculture as a *national* concern.

" Or perhaps, on producing a certificate of the bricks ufed in draining, farm culverts, &c. to the collector of excife, the brick duties may be repaid; inftead of the brick duty (which is a VERY *unequal* impoft) it has been fuggefted to lay a tax ·*per* foot on all houfes and walling of every defcription, calculated on the *mean* numbers of bricks fuppofed to be ufed, and applied to *every fort of material, always* excepting cottages and dwellings for the *labouring poor.* The exceffive brick tax is a ftrong temptation to builders to erect flight and *dangerous* edifices, which would be obviated by the above regulation."

Sect. 2.—*Of paring and burning.*

On the mofs-lands, where paring and burning is practifed, both feed time and harveft is very late, owing to the uncertainty of the weather; if wet, the burning proceeds but flowly, the feed time is confequently retarded, and the crops are by thefe means fo late as to become precarious, from the advanced feafon, being frequently expofed to frofts and fnows. If the barley from the mofs lands be well houfed, it is in high eftimation, and fetches an advanced price from the farmer, who prefers corn raifed upon thofe lands for his feed. Mr. Eccle-fton fowed one year a field of barley about the middle of June, which he houfed the following year, January 1; and this crop was all eagerly purchafed by the farmers, in the next fpring, for feed corn.

Paring

Paring and burning has been too much practised *, its destructive effects are but too apparent upon many farms where it has been frequently repeated. Great crops may have been procured, by this means, for a few years; but the soil in the end is destroyed. Upon strong bent, heath, fungous moss, matted rushes, or turfy peat lands, the practice may be good, and if only repeated till those bodies are destroyed is attended with success,

Paring, with the burning, is a laborious and troublesome mode of cultivation; its success depends upon circumstances, and one crop out of three is, in many instances, the amount of what may be expected to be reaped in security. After the sods have been dried and burned in small heaps, the ashes are spread upon the ground whilst yet warm, and the ground ploughed, sowed, and harrowed in immediately, if the weather permit. If the ashes get wet or grow cold before this operation can be effected they are injured.

Among those who have much distinguished themselves by their exertions in draining, and other improvements, James Okill, Esq. Lee Woolton, merits being noticed in the Lancashire Report. By draining and marling he has improved the value of the estate he occupies (about 60 acres, of 8 yards to the rod) to the increased amount of 30 s. per acre per annum, since the year 1780. The advance of the value of land, in this space, is, to be sure, to be taken into the account. This estate was gone over the 9th of June, 1795, and is in excellent condition. Above 1000 yards of under-draining with stone was completed in a very sufficient manner the year 1794.

At an expence of six shillings per rod of 8 yards, what a saving might have been made by Mr. Elkington's mode! Mr. Okill has filled up and drained several old marle pits, and gained, by this method only, some acres of land. It may deserve notice, that Mr. Okill stepped forward, contrary to the advice of some of his more cautious and timid neighbours, and gave excellent answers to most of the agricultural questions.

* Paring and burning is not practised in this county for green crops, but for grain.

SECT.

Sect. 3.—*Of Manuring.*

MARLE is the great article of fertilization, and the foundation of the improvements in the agriculture of this county; and this earth, or foffil, is fortunately wanting but in few places. There are feveral kinds of this article, valuable in proportion to the intrinfic quality of each, or the calcareous matter which it contains, or the nature of foil to which it is applied. To the ftiff clay lands, the blue or reddifh flate marle, full of calcareous earth, is more beneficial; but to the light fand lands, the ftrong clay marle is more genial. Thus not only a calcareous ftimulus is given, but additional matter is afforded, to correct the nature of the foils, by loofening the texture of the one, or giving adherence to the particles of the other, by the oppofite qualities of the different marles applied. Barren fand lands, and poor heaths, in the fouth of this county, have been, under the effects of marle, rendered productive, but this has been done at no fmall expence *.

Of the beneficial effects of marle let the following fact, amongft many hundreds that might be produced, ferve as a convincing proof.

There was a fandy loam land, exhaufted by repeated ploughing, under the worft fyftem of management. Major Atherton took the land under thefe circumftances into his own poffeffion. After a four years lea, and the land well dunged, he gave a coat of marle, carted the diftance of more than a mile at confiderable expence, and laid on to the amount of $7\frac{1}{2}$ rods to the acre, of eight yards to the rod. The fummer following a crop of oats was taken, and the enfuing year the ground was fpring-fallowed, dunged, and cropped with turnips, which were repeatedly hoed; after which, in 1793, five acres of the large meafure were cropped with barley, the produce of which was 552 bufhels to the maltfter, fold at 5 s. 2 d. per bufhel, befides 24 bufhels of fmall corn dreffed out, (a very fmall

---

* Improving, marling, and fencing, of Bootle marfh, coft 22 l. 14 s. 2 d *per* acre, of eight yards to the rod.

proportion

proportion to the quantity) and befides the tythe, which is an eleventh of the whole. It fhould alfo be noticed, that the bufhel by which it was meafured was the Liverpool bufhel of 36 quarts, which being reduced into Winchefter bufhels, and the tythe added, with 24 bufhels of fmall corn, the total amount would be 706 Winchefter bufhels, or 141 bufhels *per* acre, including the tythe, and value *per* acre 31 *l.*

The average produce of American lands, is faid to be ten bufhels wheat *per* acre.—*Information from Mr. Cooper, late of Manthefter.*

At Knowfley Hall, in the year 1794, 92 bufhels of wheat, of 70 lb. to the bufhel, were reaped from one acre of land, of 8 yards to the rod; after a crop of pink-eye potatoes of near 700 bufhels to the acre. Mr. Warling (the fteward) feemed to think, that if the land had been previoufly marled, the land would have given 20 bufhels *per* acre more. Sort of wheat, fouth cone.

Thefe are two rare inftances, and more than double the common average of either diftrict; but may ferve as a proof what fuperior culture is capable. Marle has been tried as a manure after being burned, which may be in a kiln after the manner of lime, or laid over a gutter, under which faggots, &c. for fuel, have been previoufly laid. It has alfo been burned in a common oven, and been found to anfwer at about ten bufhels *per* ftatute acre, after being bruifed into a kind of powder, and fown with the hand as a top dreffing. Marle is an excellent improver of the foil, under fo many different circum-ftances, that it cannot be recommended too often, nor praifed beyond its real merits. It adds to the ftaple of the foil, and improves its quality, and renders manure, of whatever kind, more effectual, with lefs in quantity; it will admit a repetition of the procefs, with equal advantage, again and again. In fhort, fo far as experience proves in Lancafhire, it feems the grand bafis whereon every agricultural improvement fhould be eftablifhed.

The fummer is the beft feafon * for laying marle upon the land,

.Where there is a dry head of marle, the winters in which there is a long froft is the propereft time for marling, as both men and horfes are le.s.

land, fometimes immediately after a crop of hay has been taken. Its effects upon the grafs are foon vifible, from the rich verdure which it produces. Long experience has fufficiently proved the propriety of the general practice of the county; which is, to lay the marle upon grafs lands—the older the better; the fward and grafs united caufes a fermentation and putrefaction, which feems neceffary to produce a proper effect.

The quantity laid on is from two to three, or three and a half, cubic roods of 64 yards to every ftatute acre; the expence, of which is, according to the diftance carried, if in the fame field, or within the diftance of fixty rods, on the average, at about eight pounds *per* acre. It is reckoned a much better practice to have the marlings repeated, with a gentle covering, than a ftrong thick coat of marle, which is intended to laft a number of years. If thefe dreffings of marle were repeated more frequently (and no hufbandry has been found to pay better), the lands in Lancafhire, in general, would be found much more productive.

The marle fhould partake both of one fummer's fun, and one winter's frofts, at leaft. After being expofed to the effects of the weather, in large lumps, it begins to fall, or melt; the particles appear unctuous and foapy, and the quality of the fubftance feems quite changed from its original ftate. Then, in the enfuing fpring, it fhould be divided (the parts now feparate with eafe), and equally diftributed upon every part of the furface, this is, with facility, effected by harrows, &c. after which it is ufually ploughed under; but, if permitted to remain a year or two longer, the lands would be more improved in the iffue, by the length of time given previous to the marle being ploughed in. But the marle does not produce its full effects upon the foil, till intermixed and incorporated by a repetition of plough-

lefs exhaufted in that cold feafon by violent exertions, and the work is done at lefs expence to the farmer, as in moft neighbourhoods, at that time of the year labourers, are moft plenty. Some few individuals lay marle pits dry, and have paces or roads ready made, in order to take advantage of a long froft.

Q.

ings,

ings, and an intermixture of dung, or other manure, for marle is not effectual without fuch addition.

This fubject cannot be too often brought under review, as from the different reports it appears fo little noticed in many parts of the kingdom. The following fact may ferve to prove, that whatever defects the Lancafhire agriculturift labours under in his general procefs, he at leaft does not labour under a poverty of fpirit.

In the year 1793, S. H. Fazakerly, Efq; of Fazakerly, purchafed nine acres of land upon Warbreck Moor, being an uncultivated part of that wretched, poor, black, fandy wafte, laid out for improvement fo long ago as the year 1761.

In the prefent year, 1794, this hitherto uncultivated lot of nine acres was marled, at the rate of nearly twelve rods, of 64 cubic yards to the acre of eight rods.

Prime coft of land, £. 33. 6 s. 8 d. per acre   -   £. 300 0 0

Marling and carting, £. 27. 15 s. 6 d. } 250 0 0
  per acre - - - - - - - }

Extra expences, with fencing - -   50 0 0
  ————————— 300 0 0

So that it appears, the expence of improvement by marle only, and before a fingle crop has been taken, amounts to the purchafe of the fee fimple of the land; befides a moft extravagant coat of dung, the expence of which is actually £. 12. 15 s. per acre.[*]

1794. Nov. 14.—Major Atherton, travelling in a chaife over Baffage Heath, on the road from Colefhill to Litchfield, obferved a wretched gravelly common under improvement by marling, a kind of flate marle. He judged the quantity laid on was four rods, of 64 yards to the acre of eight yards. The land the property of Lord Middleton.

The above is noticed, as the road is frequently travelled, to

[*] A farmer who was prefent at this calculation faid, that the real expence of dung ought to have been £. 15 per acre, Mr. Fazakerly having allowed himfelf too little in the calculation for carting fix miles. The dung laid on is cow-dung, purchafed at Liverpool at the rate of 5 s. per ton.

call

call forth the attention of the paſſenger, that the effects may be hereafter obſerved.

The quantity laid on was leſs than would have been generally given by a Lancaſhire farmer by nearly one half: it was one third leſs than the coat given by Mr. Fazakerly, as appears from the preceding ſtatement.

Anecdote.—Talking over the ſubject of marling one day in company, the following ſtory was told, which ought to be preſerved.

A Lancaſhire farmer, on obſerving the great advantage that might be obtained from the uſe of this article in a county where its uſe was not known, after ſome deliberation hired a farm, with intent to improve it by marle at his own expence. Having obtained a ſufficient length of leaſe to be reimburſed, he began the operation at the proper ſeaſon; but the practice was ſo novel in that neighbourhood, as to attract the attention of by-ſtanders, and was ſoon conveyed to the ears of the ſteward, who immediately came over to ſtop ſuch proceedings. Arguments were in vain; for what ſervice could *dirt* laid upon *dirt* prove? beſides the injury done to his lord's lands by the digging of holes, which, as a good ſervant to his maſter, it was his duty to prevent.

The ſtory concludes, as the farmer's deſigns were thus fruſtrated, he, for ſome trifling conſideration, obtained a releaſe from his contract, and left the county.

By way of contraſt, the opinion of an intelligent Lancaſhire farmer may be given, in his own words.

" Marle is a never-failing friend to moſt lands in this county; but here is a large field for improving the management of this uſeful article: the firſt and grand object is the diſpoſition of the pits. Thouſands of acres, I can ſafely ſay, are waſted, and in many places the land worſe than before. It ought to be a ſtanding rule not to ſuffer a pit to be made, unleſs it could be laid dry, which I verily believe may be done in three fourth parts of the county. One ſingle drain in many places would lay dry fifty acres, of, from 8 to 15 feet of marle a-breaſt: care ſhould likewiſe be taken, to take off the upper clay, which is

generally

generally from two to three feet.   Now this ufelefs top is excellent for the making of bricks, and when ftones are not to be had,
a coarfer kind (not paying duty) might be made to anfwer the
purpofe of draining at a very trifling expence, if nothing but
fun-dried: where a pit is laid dry no land is loft, and the farmer
may marle any feafon of the year, and befides the faving of
land, and expence in many places (according to Mr. Elkington,
which is the only true fyftem) the land below would be drained,
and many fprings cut off."

Notwithftanding there is a general propenfity to convert
arable land into pafture and meadow, as moft convenient to
the populous ftate of the county, yet an intelligent gentleman *
judicioufly obferved, that it might be occafionally neceffary to
break up grafs lands, if only for the fake of reaping the fuperior effects of marle, which not only adds to the ftaple of the
foil †, but to a certain degree improves and enriches the quality of the grafs; and a greater attention to green crops during
the procefs of the plough, would certainly afford food for a
greater quantity of ftock.   Befides, in old lays, the grafs, if for
hay, becomes too foft; if in paftures, four.   Turning over the
foil changes and improves the nature and quality of the grafs.   In

---

* J. J. Atherton, Efq. Walton-Hall. And likewife old meadow land,
that has been a deal manured. The hay will fometimes, upon fome land,
grow not fo good in quality, nor fo faleable, if for market; but by being
ploughed two or three years, more or lefs as occafion may require, will
greatly enrich the quality. Likewife an old pafture the fame, when it
can be made convenient. Land that has been exhaufted by long ploughing,
&c. and laid down poor, is generally a long time before it will come to a
proper fward of grafs, if ever fo well manured at the top, which the report
is, let it lie and manure it well, and it will do in time; which to be fure it
will. But my mode is, after it has been well manured twice, and paftured
three or four years, if it then does not do as expected, to plough it again for
a year or two, as may appear the beft, and then let it lie again, and by
fo doing it will fooner come to a fward of grafs, and the grafs will be
much richer and better; for by the manure that has been laid on the top
of the land being well mixed with the foil, and ploughed as deep as the
land will bear, it makes it much better, and lefs manure will do for the
future time, and the land much improved by it.

† A cubic rood of marle, of 64 yards to the rood, adds nearly half an
inch to the ftaple of the foil to a ftatute acre of land.

ftiff

ftiff foils, the change from arable to pafture or meadow may not be fo neceffary to be frequently repeated. But all foils improve by a judicious change of culture. Caution is however neceffary, not to yield to importunities of tenants to break up old lays, without proper reftraints.

M..rle is got by falling it in large clods; this method is expeditious, but requires great caution, and is frequently attended with danger; the piece intended to be fallen is undermined, and loofened at each fide, by being cut through; long piles are then driven in at the top, and fometimes water is required to infinuate itfelf into the interftices which the poles have made. The clod falls with fuch violence as to break the mafs into pieces.

It is no fmall confideration where to fix the pit, from whence the marle is to be obtained to moft advantage, provided there be a choice; and when there is, the following confiderations fhould be weighed: of deftroying the leaft land; of affording the leaft length of carriage, which is the heavieft part of the expence; of affording the leaft draught, by going down hill, if poffible; that the water ftagnating in the pit afterwards may not be injurious to the land and of rendering the leaft damage to the lands in future.

The

The expence of carting at different diftances may be conceived from the following rough draught here given.

Suppofe the Lot 1. be thirty rod fquare, and the pit right in the center, fo that its greateft diftance from the pit be fifteen rods every way. The cartage will be 18 s. per rod; and, to fave fractions, call the field fix acres; then the account will ftand as under:

Cartage, per rod - 18 s.
N° of acres - - 6

$$108$$

N° of rods laid } 6
per every acre }

$$648$$

£.32. 8 s. Total amount
of expenco
of Lot 1.

Now, if Lot 2 be marled out of the pit in Lot 1, the additional expence will be 12 s. per rod, or £. 54, being forty-five rods from the pit.

And, if Lot 3 be to be marled ftill from the fame pit, the additional expence will be 26 s. per rod, or 44 s. the whole; the diftance from the pit being feventy-five rod, and the expence £. 79. 4 s.

Again, Lot 4. being as large as the other three, and the pit in the center, the extreme diftance will be forty-five rods each way, and the cartage will be 21 s. per rod, of fixty-four cubical feet, and which will amount to £. 113. 8 s.

The comparative eftimate ftands thus: Lot 1, £. 32. 8 s.—Lot 2, £. 54.—Lot 3, £. 79. 4 s. Total amount of which is

£. 165.

Lot 1.
Pit.
A. R. P.
5 2 20
Lot 2.
5 2 20
Lot 3.
5 2 20
Lot 4.
as the other Lots, 2, 3.
Pit.
This Lot is as large No 1,

£.165. 12 *s.*—Three times £.32. 8 *s.* is £:97. 4 *s.*—Balance faved by having a pit in each field would be therefore £.68. 8 *s.*——The expence of Lot 4 is £. 113. 8 *s.*; and from which fubtract £. 97. 4 *s.* and the balance is £. 16. 4*s.*

The above will evidently prove the advantage of proximity to the marle; but a pit in the middle of a field is not only an eye-fore, but a nuifance; therefore, if poffible, fhould be avoided. Nor is the advantage fo great in the middle as at firft may be thought; fince, in coming out of the pit, there being only one pace, fome part of the ground muft of neceffity be gone twice over; whereas, if on one fide of the field, and Central, all the land lies immediately before the pace of the pit.

Expence of marling upon Bootle Marfh, about the year 1780, *befides fencing, &c.*

	£.	s.	d.
Getting and filling, per rod of 64 cubic yards -	0	10	0
Spreading - - - - - - -	0	2	2
Carting; the average diftance from the middle of the pit to the middle of the land, 60 rods -	1	9	0

N. B.—In this calculation there are fix carts, five in motion, each goes the diftance of twelve rods, whilft one ftands in the pit to be filled. The fize of each cart is 20,736 inches (cubical), ufually drawn by three horfes; the weight of the load abour 15 cwt. and two cubical yards of marle make about three loads.

The number of workmen are fix fillers and getters; ufually two right-handed men at one wheel, and two left-handed at the other, with one filler behind—one getter is generally fufficient.

	£.	s.	d.
Getting, filling, and fpreading, to the acre of 64 yards to the rod, on Bootle Marfh, was - -	3	19	1
Cartage - - - - - - -	9	8	0
Digging for the marle, clearing the head, expences at finifhing, &c. per acre - - - -	2	7	0
	£.15	14	1

There

There were about 6½ rods laid upon the acre on this oc-cafion.

The men got 2 *s.* 6 *d.* and the carts 7 *s.* 6 *d.* per day.

Getting and filling marle is very laborious work, and re-quires the utmoft exertion to obtain thefe wages; and this work, after all, cah only be effected by young men in their prime, cheared by the company of fellow-labourers, and fre-quent refrefhments. Five working davs are reckoned equal to fix, for they ufually begin at half paft four in the morning, and reft one hour at breakfaft, from eight to nine; reft again from twelve till two, and then work till fix; and generally get out nine rods per week.

	£.	s.	d.
The prefent price is—			
For getting and filling, per rod - - -	0	12	0
Spreading - - - - - - -	0	2	6
Carting - - - - - - -	1	13	0*

ADDITIONAL INFORMATION ON MARLE.

Marle is the foundation of all improvements in the agri-culture of this county; and here the hufbandmen of Lan-cafhire and Chefhire may afford an ufeful leffon to the reft of the kingdom: fo well are they convinced of the neceffity of attending to this primary object, that neither labour nor ex-pence deter them from the moft vigorous application of it. There are feveral varieties of this foffil manure valuable in proportion to its intrinfic qualities, or the nature of the land to which it is to be applied. Shell marle or flate marle are more defirable in the ftiffer and more clayey diftricts, inaf-much as they contain a large proportion both of calcareous matter and of fand—clay marle in an inverfe ratio more ge-

* This fubject has been detailed to a greater length than fome may think requifite; but marling is in this county performed in a mafterly manner. The particulars here collected may be ufeful, on future occa-fions, to the farmer, as the documents are only regiftered in the memory of old practitioners. It is with no fmall difficulty that the feveral *data* are fometimes obtained and afcertained, and it was with fome labour they were collected for the prefent purpofe.

*

nial to a light and fandy diftrict, as in both thefe circumftances the natural defects of the foil are in fome meafure obviated. Undoubtedly the calcareous matter contained in either marle is of the higheft importance; but obviating the natural deficiencies of the foil, by adding fand to clay or clay to fand, is of more confequence than the mere calcareous ftimulus, which might be obtained at a much lighter expence. Innumerable inftances are to be found in this part of Lancafhire, where barren heaths and wretched fands of all defcriptions have been rendered in the higheft degree productive b this admirable foffil; indeed there is reafon to believe that by far the greateft part of the diftrict has been reclaimed by marle. The great confequence of making fuch a practice more generally known need not be expatiated upon. It is of the utmoft importance to attend to the *application* of marle. The general cuftom is to lay upon the great Chefhire acre, of eight yards to the rood, from three to feven roods, of fixty-four fquare yards each. From four to five rood may be confidered the average quantity to the acre (one Chefhire acre contains two acres and eighteen perches and a half of the ftatute meafure) more and lefs are frequently applied, but the quantity ought indifputably to be in proportion to the quality of the foil and quality of the foffil. The general experience of this country has proved to a demonftration the propriety of its univerfal practice, *viz.* to lay it upon grafs land which is intended to be broke up the enfuing fpring. This fyftem is however carried by fome of the old farmers to an abfurd length, as they will not made any land, however neceffary fuch an operation may be, unlefs it has been a given number of years under grafs. Sometimes what is provincially called a *coat* of marle has been fpread upon the green fward, and left unploughed many years; in this cafe the grafs fometimes receives confiderable detriment, as the marle finks downwards in a body without incorporating with the foil; though when marle has lain feveral years in this ftate, the fubfequent crops of corn have been found to be enormous.

The general rule is to begin marling about May or June, in fhort when fpring feedings are over, continuing as oppor-

R   tunity

tunity ferves throughout the fummer months; it is not, however, unufual to take a crop of hay before the marling is begun; in either inftance, the effects of the marle become
fpeedily vifible by the rich verdure of the grafs, which affords
a pafturage of the moft beneficial nature.   Marle is fpread immediately after carting, but left in a rough lumpy form, that it
may be expofed as much as poffible to the viciffitudes of the feafons; if it contains a large proportion of clay it will remain for
many weeks, perhaps months, in large unwieldy lumps, though
in general the rains of the latter end of autumn, and the fucceeding frofts of winter reduces it into the form of an unctuous but friable material, the further difperfion of which is
eafily effected with clotting beetles, fpades, or harrows; this
difperfion however ought not to be attempted till a week or a
fortnight before ploughing, as the moft beneficial effects are
produced by alternate rains and frofts; and by this long expofure it is more than probable that the foffil   may acquire by
attraction the moft nutritive qualities :—the turf, when ploughed under, anfwers the purpofe of a rich vegetable manure.

So far the Lancafhire farmers have confiderable merit, but
their fubfequent conduct deferves the higheft cenfure; many
of them taking repeated crops of oats with the interval of one
fummer fallow for wheat, by way of cleanfing the land; after
which barley and oats again, as long as the land will produce
any thing; and then laid down again, as ufual, with weeds and
couch-grafs.—The courfe I fhould recommend would be, to
take one crop of oats the fpring fubfequent to the marling—
plough the ftubble immediately, in order to expofe the marle
again to the influence of the froft—fallow with manure for
turnips, a crop that under this management is never known
to fail—then barley, clover, wheat, turnips fed off with fheep,
and barley again, with well-dreffed hay-feeds, and white clover
and trefoil, for a perennial lay, or at leaft for fome years.—
Land thus hufbanded produces in a moft exuberant degree,
and at the fame time is rendered perfectly clean from all
weeds, without being in the leaft haraffed.   Poor fandy foils
are thus rendered capable of producing a covering of the rich-

---

* See Kirwan's Differtation on Manures, An. Ag. vol. XXIII. p. 105.

eft graffes, and under proper management may be depended upon in all feafons. If this procefs of marling is again repeated, the nature of the land is totally altered, nor is it probable that any ftranger to marle will give credit to the immenfe benefit to be derived from fuch management, without ocular demonftration.

Some time ago I gave the furveyor of this diftrict, Mr. Holt, the particulars of a field of eight acres marled by me in 1790; as I underftand he has fent an account of it to the Board of Agriculture, I cannot but add the following particulars—that it had been ploughed and haraffed in the moft barbarous manner for many years previous to my getting poffeffion of it, the out-going tenant paying about £. 2. per acre for it, and under no reftrictions—that fome years ago one of the beft acres in the field was manured for barley, which did not produce more than one quarter in return, the crop being moftly deftroyed by weeds, and the land incapable of nourifhing grain—that when in grafs it was frequently unnt for mowing before September, and then the grafs not more than fix inches long—that in 1786-7, when I took poffeffion, it was manured at the rate of 50 tons per acre of horfe and cow muck and night-foils from Liverpool—that it was attempted to be paftured from that time till 1790, but that the cattle were all always difcontented with the grafs, and the land was returning by degrees to its original ftate of moor—In 1790 I marled it at the rate of feven roods and a half per acre—the expence was very great, as I carted the marle near a mile and a quarter.

		£.	s.	d.
1791.—Oats, 720 meafures, of 36 Winchefter quarts each, at 3 s. the meafure - -		108	0	0
1792.—Turnips manured, and twice hoed— drawn for cattle:—paid the expences of the manure, the hoeing, &c. &c.				
1793.—5 acres barley produced 576 meafures, of which 552 meafures fold to a maltman at 5 s. 2 d. - - - -		142	12	0
3 acres oats, 330 meafures, at 3 s. -		49	10	0
1794.—Clover, two excellent crops, and now in wheat.				

R 2   March

March 1795.—I fent 4 fpecimens of the marle from the fame pit the above-mentioned eight acres were marled from, to Mr. M. Renwick, Chymift, in Liverpool, to be analyfed: the products were, from 100 grains of each:

	N° 1.	N° 2.	N° 3.	N° 4.
Flinty Sand - - Gr.	40½	40¼	34	32 7/10
Clay and filicious Earth -	39½	·39	44	47 3/10
Calx - - - - - - -	19¼	20½	22	20
Loft - - - - - - -	9¼			
Gr.	100	100	100	100

This average, however, is not ftrictly juft, becaufe the depth of the ftrata were extremely unequal:—of 14 feet, probably there were, of N° 1, 2, and 4, not more than 1 foot each; the remainder of N° 3, which is beyond all comparifon the beft marle in itfelf, and beft adapted to a deep loomy fand.

The reafon the Leicefterfhire farmers object to marling, is, that it is inimical to grafs,—in Lancafhire we know we can get no good grafs without it. What is the reafon of this difference of opinion? It arifes from the application; if marle, according to the Leicefterfhire fyftem, is put upon a turnip fallow, and immediately ploughed under in its crude ftate, no wonder it produces no grafs and little corn.—In Lancafhire, it is expofed at leaft five months, and always to a winter's froft: and hence arifes the benefit to grafs. The fheep walks and rabbit warrens of Norfolk were reclaimed by marle; the original marlers and their fucceffors grew rich, and the land

9   produced

produced exuberantly: at prefent we are told, that the land will not bear marling twice (an *idea* utterly difcountenanced by the *practice* of this country); and from this mifchievous idea it arifes, that the county of Norfolk is faid to want another fyftem of cropping, that turnips and clover come round too often, &c. &c. (See Kent's Survey). What are we to infer from this? that the fame material which caufed the original improvement will reftore it to its former fertility—at any rate the experiment is worth trying.

During the late hard froft, 1794-5, I have been marling about 4¼ ftatute acres of fandy loam, in order to keep my labourers employed, who muft otherwife have applied to the parifh for affiftance. I carted the marle near a mile in the large three horfe carts of this country, which I fhould conceive brought about 26 cwt, at a load.

I find, in the Ann. Agr. Vol. XIX. p. 476. that Mr. Colhoun performs his claying in carts that cube 35 feet; my carts cube about 21 feet; of thefe 175 loads bring about two roods the ftatute acre,

$$175 \times 21 = 3,675$$
$$105 \times 35 = 3,675$$

But Mr. Colhoun prefers 80 loads per acre on a blowing fand; that is, 1,400 loads to 2 Chefhire acres—

$$2 \text{ Chefhire acres} = 4 \ 0 \ 37 \text{ ftat.}$$

I bring 350 loads to a fandy loam. This matter merits the attention of the Board.

Herefordfhire, which is *all marle*, produces from 2 qrs. 4 b. to 3 qrs. 1 b. per acre of wheat; about 4 qrs. 5 b. of barley; about 3 qrs. of beans; peas the fame *.

* Clark's Report, Heref. p. 26.

Corfe

Corfe Lawn, and the Foreft of Dean, in Gloucefterfhire, are all marle. The clay pits of Feverfham, and the brick earth of Kent, in general, are marle. The clays of Nottingbamfhire contain excellent marle. The fineft marle I ever faw in England is dug out of the coal road near Mr. Tate's houfe by Loughborough. Mr. Bakewell has marle upon his farm. What are we to think of the abfurdity of carrying Norwich marle 40 miles by water, to a fandy country abounding with a clay, that in the open air breaks into fmall die-like pieces, and which, upon analyfis, yields from 100 grains

Of impalpable clay	—	—	50 grains.
Of fand	—	—	7
Calx	—	—	43
			100 grains [*].

Sea flutch, from the Ribble and Wyre, is, in fome places adjacent, made ufe of as a fubftitute for marle, to which it is reckoned equal, but in general not fo durable [†]. It is frequently ufed as a fubftratum for fruit trees at Formby. The quantity is a load to each tree; its effects are wonderful. This practice, however, may not prove beneficial where the foil is denfe. At Roffal in the Filde, where there is no marle, after a ftratum of ftrong clay under the foil, they pafs through a fand with cankered veins, next a fand with fky-blue veins, with thin fhells like barnacles, called in the provincial phrafe hen-fifh; and this proves a good fubftitute for marle. Sea-flutch, particularly at Wefton, a village in Chefhire, near Frodfham, is found to be much more fertilizing and more permanent than marle, I mean that part of a falt marfh which has been graffed over for a few years; for that which is overflowed daily contains more fand, and is lefs enriching. I do not think that all

See Marfhall's Norfolk, Vol. I. p. 24. and Vol. II. p. 193.

[†] Mr. Standen, fteward to Bold Fleetwood Hefketh, Efq. fays, more durable than marle.

the

the manures in ufe in the kingdom combined could have a better effect either upon arable or grafs land than the above-mentioned foil has. This manure, where the plough is not immoderately ufed, will laft thirty years. It is ufed in much the fame manner and quantity as marle.

Befides the dung got from the farm-yards, there are great quantities raifed by the cowkeepers and ftablekeepers in the large towns. At Liverpool horfe-dung fells at about 5 *s.* 6 *d. per* ton, cow dung from 4 *s.* 6 *d* to 5 *s.* 6 *d. per* ton, butchers' dung 6 *s. per* ton, the afhes mixed with privies, fcraping of the ftreets, &c. under the denomination of night foil, about 2 *s.* 1 *d. per* ton *, Liverpool alfo occafionally has the dregs of blub-ber from the whale fifhery after boiling the oil, which mixed with foil, is a rich manure, but not lafting. Soap afhes alfo, if put upon old lays, have been found very advantageous, and very durable in paftures, but not fo durable either on ploughed land or in meadow †. Soap afhes, like lime, do not at all an-fwer upon a black foil with fand underneath, neither have they their immediate effects, like dung upon any foil, but are very fertilizing and durable after the firft year, when applied upon dry pafture or meadow lands if they be fox foils, and very much change the nature of the graffes; viz. from very indifferent forts to wild clover, trefoil, &c.—Rape duft has been found to anfwer, laid on at about fixty bufhels to the acre, and cofts about 10 *d. per* bufhel ‡. Soot is alfo ufed in the fpring, and thrown with the hand upon the corn; this is often practifed upon poor exhaufted lands, and, if rain immediately enfue, with fuccefs; but there feems fomething at prefent in-explicable about the proper application of lime, or its operation upon different foils. It has been frequently tried without any apparent utility, and it fhould appear that lime requires fome

---

* At Manchefter, cow and horfe dung are about 1 *s. per* ton higher.

† Quantity 40 to 50 ton *per* acre, from 8 *s.* to 10 *s. per* ton at Liver-pool.

‡ There muft be an error in the price of rapeduft, as it is no where to be had under 2 *s. per* bufhel, and fome places 2 *s.* 6 *d.* it is a very ufeful tillage upon cold lands, efpecially upon meadows, but not durable.

particular

particular fubftance in the foil whereon to act, to produce any good effect. Lime has in general not been found to anfwer fo well a fecond time as at the firft operation. It alfo requires a fward, or vegetable roots, to produce fertility *; and it more frequently fucceeds when mixed properly with earth either on fallows or fwards. *Lime* is the beft manure for grafs lands, either laid on by itfelf or in compoft, if ufed in *fufficient* quantities In a farm of a cold clay foil, after draining near twenty years, the *lime* was laid on the fward in *May and June* to the amount of two hundred bufhels on a *ftatute* acre; the lands have not been ploughed, but have yielded the *fineft grafs* for hay and pafture, and yet appear to be in a ftate of improvement. The ufe of lime as a manure has nearly fuperceded that of *marle*. Immenfe quantities of lime-ftone are brought by the rivers and canals from *Wales*. Great was our alarm laft year when the tax on *ftone* was propofed, but although the fixth fection of the 34th Geo. III. c. 51, exempts our *lime-ftone* imported from the tax, yet the requiring the *ufual coaft difpatches*, and certificates for the floops employed in this trade, occafion great expence and delay, and operate as an heavy and unneceffary impoft, without any advantage whatever to the revenue. This is an overfight eafy to be remedied, and from good authority, I learn has in a great meafure *fruftrated* the wife intentions of the legiflature relative to the taking off the duties on *coal* and *falt* exported coaftwife in *Scotland*, and calls for the *efpecial* attention of the reprefentatives of *North* Britain.

But neither marle nor lime produces any good effects upon the exhaufted lands of the Filde, which have undergone the *centennial ordeal*. Upon thefe occafions the farm-yard dung feems to be principally wanted, to reftore the oily part extracted by fuch a continued fucceffion of exhaufting crops.. So great a quantity of land is ploughed without a proper rotation of green crops, for the ftock which ought to be kept, there is no refource for raifing dung but from the cattle, as there are no towns fufficiently large to afford proper affiftance, nor yet canals to bring it from diftant places.

* There are fome exceptions, neverthelefs, even to this.

In

In the Leicefter Report it is mentioned, that fome perfon had burned bones with lime for manure.   In November 1794, the furveyor mixed about equal quantities of bones with lime (of the latter what is fold at Liverpool for 40 meafures, at 8 *d. per* meafure) in different ftrata, and clofely covered with fine mould, well clofed with the fpade.   This heap continued burning about ten days, after which the whole was mixed together with earth, the bones being chiefly reduced into fmall parts, or if the parts yet adhered, flew afunder with a fmart ftroke.   It is imagined that the moft effential or oily part of the bone would efcape with the fmoke during this procefs, the fmoke being great and the fmell fœtid.   Not having convenience to bruife the bones, this method was adopted by way of experiment upon the hint given.

Bone-duft, or bones ground in a mill, have been ufed with fuccefs by William Mayor *, the farmer at Afhworth-hall, near Rochdale.   He has two fluted iron rollers placed at the end of a corn-mill fhaft, which grinds them expeditioufly ; he applies them to his own grounds, and difpofes of them to different purchafers.   Near the fea good compofts have fometimes been made of land-lime, earth, dung, and fea weeds, with a fpecies of fhell-fifh growing upon the rocks, which is found to be an excellent manure for barley.   The fcrapings from the ftreets, along with afhes and night foil, have by an experienced farmer † been mixed together with lime in the following proportion : to every twenty tons weight of this black muck (as it is fometimes called) he adds about forty bufhels of lime, which he mixes together before the lime runs to mortar (his own expreffion) which deftroys the good effects, and prevents a proper incorporation, and which anfwers well upon either dry or wet lands, particularly when laid down to either' paftures or meadows.   The drainings from the farm-yard have been of late, by fome good farmers, collected into one place, and, if they cannot be thrown over the lands any other way, are conveyed in cafks by carts, and diftributed upon the land by means of a trough perforated with holes.

* And by George Clayton, Efq. Loftock, near Prefton.
† Mr. Henry Harper, Bank Hall.

S   The

The skimmings of sugar under refinement, when boiling, is a rich manure; so much so, as to take three parts of soil to mix together. Three loads of earth, and one load of these skimmings, which consists of American clay and other fertile ingredients, make four loads of rich and durable manure.

Experiments on Manures, by Mr. Henry Harper.

" The following experiments of different kinds of manure will shew the difference of both quantity and the quality of produce on the different kinds of land on my farm, on which I manured half an acre of eight yards to the rod with every kind of the following manures; and when made into hay, as nearly all alike as possible, I weighed one average square rod from every lot.

Lot the 1st.—Horse, cow, and butchers dung, all mixed together, of each about an equal quantity, which lay in that state about two months, and then turned it over, and let it lie eight or ten days, and then put it on the land before it had done fermenting, and spread it immediately. This was set on in September 1793.—The produce 3 stone 15 pound per rod, at 20 pounds to the stone.

Lot the 2d.—Horse and cow dung, mixed and turned over the same as Lot the 1st, and set and spread on the land at the same time.—Produce 3 stone 14 pound per rod.

Lot the 3d.—Horse dung, turned over and set on the land the same as Lot the 1st.—Produce 3 stone 13 pound 8 ounces per rod.

Lot the 4th.—Cow dung, turned over and set on the land the same as Lot the 1st.—Produce 3 stone 13 pound 8 ounces per rod.

Lot the 5th.—Night-soil, coal-ashes, and cleaning of the streets, and about 40 measures of lime to every ton weight, and turned over while the lime was in its floury state, and not suffered to run to mortar, for then it is of little benefit; one part of this was set on in September 1793, the other part the middle of March 1794, but no difference in the crop to be perceived.—Produce 3 stone 13 pound per rod.

Lot

Lot the 6th.—Night-foil, coal-afhes, and cleaning of ftreets, fet on the land in the fame manner and times aś Lot the 5th, and no difference in the cropping part.—Produce 3 ftone 2 pounds 8 ounces per rod.

Lot the 7th.—Marle frefh got, and mixed with an equal quantity of horfe and cow dung, and lay about three months and then turned over, and lay a month and then turned over again, and put on the land in fix or eight days, and at the fame different times as the two laft lots, but no difference in the cropping.—Produce 3 ftone 8 pound 12 ounces per rod.

Lot the 8th.—Water from a refervoir that all the urine from the ftables, cow-houfes, and all drainings from the dung-hills, farm-yard, hog-ftyes, and all the wafte water from the houfe runs into, and is carried on the land in a watering-cart made on purpofe that holds four hundred gallons ; and the water was put on the land in April, about 12,000 gallons to the acre of 8 yards to the rod, and again in May 12,000 more. —Produce 3 ftone 5 pound per rod.

Lot the 9th.—Blubber, the offal of whale-oil, mixed with foil, and fet on the land the 1ft of April 1794.—Produce 3 ftone 2 pound 8 ounces per rod.

Lot the 10th.—Soot, fowed on the land the middle of April 1794.—Produce 3 ftone 1 pound per rod.

Lot the 11th.—Plafter of Paris (gypfum) fowed on the land in April, the weather then fhowery and favourable for it. —Produce 2 ftone 2 pound per rod.

Lot the 12th.—No manure at all.—Produce 2 ftone 2 pound per rod: fo much for gypfum, that has been made fuch account of.

Lot the 13th.—Soap-afhes or muck, fet on in March 1794. —Produce 2 ftone 10 pound per rod.

Lot the 14th.—Lime, fet on in March, clean by itfelf.—Produce 2 ftone 8 pound per rod.

An improvement by way of experiment upon Lots the 1ft, 2d, 3d, 4th, and 5th, water from the refervoir put on thefe lots the beginning of May 1794, at the rate of 12,000 gallons per acre.—Produce 4 ftone 8 pound per rod.

Lot

Lot the 1ft.—Produce 3 ft. 15 lb. per rod, is 600 ftone per
              £. s. d.

		£	s	d
Horfe, cow, and butchers dung.	acre, at 5½ d. per ftone —	13	15	0
	After-grafs, per acre —	2	2	0
		15	17	0
	Manure 30 tons, at 5 s. per ton —	7	10	0
	Balance in favour of the farm —	8	7	0

Lot the 2d.—Produce 3 ft. 14 lb. per rod, is 592 ftone per

		£	s	d
Horfe and cow dung.	acre, at 5½ d. per ftone —	13	11	4
	After-grafs, per acre —	2	2	0
		15	13	4
	Manure 30 tons, at 5 s. per ton —	7	10	0
	Balance in favour of the farm —	8	3	4

Lot the 3d.—Produce 3 ft. 13 lb. 8 oz. per rod, is 588 ftone

		£	s	d
Horfe dung.	per acre, at 5½ d. per ftone —	13	9 ·	6
	After-grafs, per acre —	2	2	0
		15	11	6
	Manure 30 tons, at 5 s. per ton —	7	10	0
	Balance in favour of the farm —	8	1	6

Lot the 4th.—Produce 3 ft. 13 lb. 8 oz. per rod, is 588 ftone

		£	s	d
Cow dung.	per acre, at 5¼ d. per ftone —	13	9	6
	After-grafs, per acre —	2	2	0
		16	11	6
	Manure 30 tons, at 5 s. per ton —	7	10	0
	Balance in favour of the farm —	8	1	6

Lot the 5th.—Produce 3 ft. 13 lb. per rod is 584 ftone per

		£	s	d
Night-foil, coal-afhes, cleaning of the ftreets, and lime.	acre, at 5½ d. per ftone —	13	7	8
	After-grafs, per acre —	2	0	0
		15	7	8
	Manure 30 tons, at 4 s. 6 d. per ton —	6	15	0
	Balance in favour of the farm —	8	12	8

Lot the 6th.—Produce 3 ft. 2 lb. 8 oz. per rod, is 500 ftone

		£	s	d
Night-foil, coal-afhes, cleaning of the ftreets.	per acre, at 5¼ d. per ftone —	11	9	2
	After-grafs, per acre —	1	11	6
		13	0	8
	Manure 45 tons, at 2 s. per ton —	4	10	0
	Balance in favour of the farm —	8	10	8

Lot the 7th.—Produce 3 ft. 8 lb. 12 oz. per rod, is 550

	£.	s.	d.
Marle, horfe and cow dung. } ftone per acre, at 5½ d. per ftone —	12	12	1
After-grafs per acre — —	1	11	6
	14	3	7
Manure 45 tons, at 2 s. 6 d. per ton —	5	12	6
Balance in favour of the farm — —	8	11	1

Lot the 8th.—Produce 3 ft. 5 lb. per rod, is 520 ftone per

Water from refervoir. } acre, at 5½ d. per ftone —	11	18	4
After-grafs, per acre —	1	11	6
	13	9	10
No expence for manure only labour, and that not fo much as the other manures.			
Balance in favour of the farm — —	13	9	10

Lot the 9th.—Produce 3 ft. 2 lb. 8 oz. per rod, is 503 per

Blubber and Oil. } acre, at 4 d. per ftone — —	8	7	8
After-grafs per acre —	0	15	0
	9	2	8
Manure, the expence per acre —	3	0	0
Balance in favour of the farm —	6	2	8

Lot the 10th.—Produce 3 ft. 1 lb. per rod, is 488 ftone per

Soot. } acre, at 4 d. per ftone —	8	6	0
After-grafs, per acre —	0	15	0
	9	1	0
Manure, expence per acre —	2	10	0
Balance in favour of the farm — —	6	11	0

Lot the 11th.—Produce 2 ft. 2 lb. per rod, is 336 ftone per

Gypfum. } acre, at 5½ d. per ftone —	7	14	0
After-grafs, per acre —	1	5	0
	8	19	0
Manure, expence per acre —	2	10	0
Balance in favour of the farm —	6	9	0

Lot the 12th.—Produce 2 ft. 2 lb. per rod, is 336 ftone per

No manure. } acre, at 5½ d. per ftone —	7	14	0
After grafs per acre — —	1	5	0
	8	19	0
No expence for manure.			
Balance in favour of the farm —	8	19	0

Lot

Lot the 13th.—Produce 2 ft. 10 lb. per rod, is 400 ftone per   £. s. d.

Soap-afhes.   {   acre, at 5½ d. per ftone   —   —   9 3 4

     After-grafs, per acre   —   —   1 10 0

                 10 13 4

Manure 16 ton of foap-afhes, at 9 s. per ton -   7 4 0

Balance in favour of the farm   —   —   3 9 4

Lot the 14th.—Produce 2 ft. 8 lb. per rod, is 384 ftone per

Lime.   {   acre, at 5½ d. per ftone   —   —   8 16 0

    After-grafs, per acre   —   —   1 10 0

                 10 6 0

Manure, 12 fcore meafures of lime, at 13 s. 4 d.

  per meafure   —   —   —   8 0 0

Balance in favour of the farm   —   —   2 6 0

Experiments upon Lots the 1ft, 2d, 3d, 4th, and 5th.—Produce of all thefe five lots equal, 4 ft. 8 lb. per rod, is 704 ftone per acre, at 5½ d. per ftone, is   —   —   16 2 8

After-grafs on all the five lots equal   —   —   2 5 0

                 18 7 8

Amount of produce of Lot the 1ft   —   —   15 17 —

Balance in favour of the refervoir-water   —   —   2 10 8

Now thefe lots are all in one field, which is old meadow land all of one quality, the foil 11 inches deep, and a ftrong loam betwixt fand and clay with a reddifh caft, and is what I call fox-land; and under the foil is a black loam fand fix inches deep, and then marle of four yards deep, and bottoms on a red fand.

This field is not to be confidered as a poor worn-out field, but has been regularly manured every third year; which if it was worn out and kept poor it would not produce one-third part of neither hay nor after-grafs, which I daily fee on fome adjoining land of the fame quality, for which I take one-third part of the value of Lot the 4th, both of hay and after-grafs, which is £. 5. 3 s. 10 d.

Lot

	£.	s.	d.
Lot the 4th.—The amount of hay and after-grafs, at £. 15. 11 s. 6 d. per acre for three years, is — — —	46	14	6
Difcount for three years manure —	7	10	0

| Balance to the farm — — — | 39 | 4 | 6 |

	£.	s.	d.
The amount of hay and after-grafs for one year, without manure, is £. 5. 3 s. 10 d. which fay three times — —	15	11	6

| Balance in favour of manure — — | 23 | 13 | 0 |

Now it is to be confidered that the farm fup-ports the expence of all labour and other expence : but to fhew the neat balance in favour of manure, fay, the manure brought to this field on my farm is four fhillings for every thirty hundred weight —

Thirty ton twenty load, at 4 s. per load, is — — — £.4 0 0			
To fpreading on the land, allow-ance, &c. — — 0 7 6			
	4	7	6

| Clear balance to the farm for three years in favour of manure, for one acre £. 19 | 5 | 6 |

I approve moft of the manure the five firft lots are manured with, although it comes higher; they require the leaft labour, which moftly pays the beft in the end, although it appears that fome of the other lots afford more clear profit; but the moft profit comes from that manure that continues its ftrength the longeft in the land.

The moft clear profit I experience is from lot the 8th, wa-ter from the refervoir, which is no coft but labour, and that not fo much as any other kind of manure; but it will not an-fwer put on in hot dry weather, for it burns up all before it,

8   except

except it was to be kept conftantly wet, of which the fupply is moftly fcarce at that time.

Lots the 9th and 10th.—Blubber and foot I would not put on land for meadowing upon any condition, for the hay is bad; and, by a conftant ufe of them, they exhauft the land, fo that it won't produce any thing at all; and they are only manures for juft the crop, with little or no after-grafs.

Soot is good for wheat, and other fpring corn, if it is fown in fhowery weather.

Lot the 11th.—Gypfum is of no ufe on my farm, neither for corn nor grafs.

Lots the 13th and 14th.—Soap afhes and lime, they do not anfwer on my farm; they keep me too long out of the profit. What they might do in time, I have not experienced; but I always think the quickeft return pays the beft, fo that the manure is not exhaufting to the land.

The water from the refervoir paid not amifs, which was fet on the five firft lots, which was an equal improvement of 2 *l*. 10 *s*. 8 *d*. per acre; and if the extra labour was to be charged, it would be a difcount of fifteen fhillings, which would reduce it to 1 *l*. 15 *s*. 8 *d*. clear profit per acre.

Now, to try the quality of all the lots, I put a fmall handful from every lot in a dry clean place, where there was little or no grafs, and they were laid promifcuoufly down, and regularly marked and numbered, to avoid miftake. And I had for the experiment fix horfes up in the ftable, all well fed with clover frefh cut: and I turned one out, and let him go of himfelf amongft the lots promifcuoufly, and when he got amongft them, fome he fmelled at, and others he tafted (there were 19 different lots); the firft lot that he fettled at was N° 8, and he eat it all clean up; and he then fauntered about as before, and got to lot the 5th, and eat it all clean up; and then fauntered as before, and got to lot the 7th, and eat it all clean up; he then fauntered as before he had done, and fmelled, and tafted, and went off from amongft them. I then put him up, and turned another out, which did exactly in the fame manner as the firft had done.—— *N. B.* And he then fixed upon the fame lots as the firft horfe had done, which were immediately taken away with
care,

care, fo as not to difturb the horfe, which through the whole of the lots were always replaced with the fame kind of hay; and out of the whole fix horfes there was little or no variation, for the next horfe that came out always fixed on the fame lots as the laft had eaten up, after being replaced. ---- And he then fixed upon lot 8, as the firft horfe had done, and eat it all clean up; and then upon another of the fame; and continued till he had eaten four out of the five, and then went off from amongft them. I then put him up, and turned another out, and he did as the others had done, and fixed upon the firft lot, N° 8, and eat it all clean up; and then to lots the third and fourth, which he eat all up, and then fauntered off. I then put him up, and turred another out, which did exactly the fame as the others had done, and fixed upon lot the 2d, and eat it up; and then he fixed upon lot the 11th, and eat it up; and then he fixed upon lot the 6th, and eat it up; and he then went off. And I put him up, and turned another out, which did exactly the fame as the others had done; and he fixed upon the laft expe-rimental lot, and eat it all up; and then to lot the 13th, and eat it all up; and then to lot the 14th, and eat it all up; and he then went off. And I put him up, and then turned the laft horfe out, which did exactly the fame as the others had done, and juft tafted of lot the 12th; but the 9th and 10th lots ftill remained, and never a horfe out of the number of fix tafted of them, only fmelled at them. And I then turned them all out together, and they made to where the lots had been, and eat up the remains of lot the 12th; but they all went off and left the 9th and 10th lots unnoticed.

And I ftill let them remain in their places till the cows came up in the evening, and never a cow, out of thirty, tafted of them (9th and 10th lots); they fmelled, and even bellowed and roared, and fcraped with their feet, and flung it about with their horns.

Now I will leave it to every reader to judge for himfelf, which of the lots were of the beft quality, and the moft nu-tritive; for myfelf, I prefer thofe that were eaten the firft.

The before-mentioned ftatements fully prove the profits arifing from manuring, to that of letting it lie by in a ftate to

produce

produce what it will do with little or no improvement, and the land to be equal as good in quality.

But the difference of improving good land, and land of inferior quality, differs from five per cent. to a hundred per cent. in its return of produce; but there is little or no land but what will anfwer in fome degree of profit towards the encouragement of improving, except fome barren mountains, and low boggy land, that lies fo low that there is no fall for draining to be come at; but the naked eye is often deceived, whence I fhould recommend a level, for this kind of land mostly anfwers the best if it can be laid dry.

There is a great difference in different kinds of manures, anfwering in different diftricts; and there are feveral in this diftrict that have not been brought to trial, fuch as burnt marle, &c. I have an experiment of it now on trial, which, as a top manure, by hand, appears to anfwer every expectation for grafs land.

The mode I prepare it is, to get the marle quite frefh out of the pit, not to take any that has been expofed to the weather, and burn it in a brick oven; and when burnt through, draw it out, and pound it into duft, which is done with little labour; and then fow it on the land quite dry, at the rate of about 15 meafures to the ftatute acre; and the expence of fire, and all trouble attending it, is eight pence per meafure.

But this is a new experiment with me, and I have only tried it this latter end of the year 1794; I mean to proceed with it for frefh experiments for different crops of both corn and grafs this next year 1795; and as foon as I have proved its qualities, I mean to explain them to the public.

I manured the fame quantity of lots of land that had lain two years, for a third crop of hay, with the fame kinds of manure as the firft mentioned lots were manured with, and with the fame experiments in regard to the quality, which nearly anfwered every defcription of the before-mentioned lots, only feventy ftone lefs in quantity per acre, of eight yards to the rod.

The mode of burning marle above alluded to, and tne ufing it when burnt as a top dreffing, is particularly recommended to the attention of the Norfolk farmers.

COAL

COAL ASHES.

This fort of manure has been known to kill rushes. The furveyor has tried their effects this year (1795), not with total, but fome effect. The beft cure, without doubt, is draining the land. Take away the pabulum, and the plant perishes.

LINSEED.

Dr. Ormerod of Rochdale made the following trial, which he thus relates :

" With linfeed flour (it is linfeed ground to powder) this I ftrewed on meadow ground, but fo lately that I cannot perceive any evident difference in my crop of grafs, but, perhaps it may anfwer in the fog; it appears to do well to corn, and to pafture ground, and I find the cattle extremely fond to eat where it has been, but fear the expence of it will prevent it being of any fervice to the fociety, as it coft me £. 1. 1 *s*. for every cwt. I ufed. I have alfo been trying the blue and white foap manure, which anfwers well to corn, but not better than afhes with necelfary manure.

" I have been using lime in a variety of forms, both next the foil upon black manure, and mixed intimately with mould and laid up together."

SEASON OF LAYING ON DUNG.

Much has been frequently faid on the beft feafon of laying dung upon the lands—the furveyor has been favoured with the following obfervations on this fubject, by an experienced farmer

" If cow-dung, the frefher the better, provided it be the proper feafon for putting it upon the land; which is, if meadow, from the time of getting the hay off the land, till the middle of October. For, if the grafs has done fpringing, the dung lies expofed all the winter to rain, fnow, frofts, and the viciffitudes

Mr. Henry Harper, Bank Hall.

of

of feafons, which exhauft the ftrength, fo as to deftroy much
of its good qualities: if it cannot be accomplifhed in autumn,
then the enfuing fpring; and if the feafon fhould not fuit, the
ftrength of the manure will be reaped the enfuing crop."—He
recommends turning over the dung previous to its being put
upon the land, and to lie till it begins to ferment; then to
carry it upon the land, and even fpread it before the heat be
gone off, and by which the dung *takes* to the land the better.
He prefers mixing cow-dung, horfe-dung, butchers-dung, and
night-foil, together, in preference to each feparate; and this
mixture is in its beft ftate from fix to eight months old.

Sect. 4.—*Weeding.*

A good crop of grain and weeds cannot exift together;
therefore, in order to fecure the former, if the latter abound,
they muft either be eradicated, or the crop greatly injured.
Except in the potatoe culture, and what little has been done in
hoeing turnips, hand weeding is in general alone practifed.
Fallows are introduced to kill weeds, where the lands are foul
by ill management. When lands have been full of couch-
grafs, a crop of hemp upon fuch lands, if well dunged, has
proved an effectual remedy *. But at prefent, there is very
little of either hemp or flax fown in the county.

There are many flovens, who too much neglect clearing their
foul crops; and many are as remarkable for their great attention,
and employ both women and children to hand weed the corn,
when about fix inches high. Many paftures and meadows are
carefully overlooked that no dock, &c. appear. Mr. Bailey's
eftate of Hope is a fpecimen of cleanlinefs.

Sect. 5.—*Watering.*

Watering Lands is much neglected in this as well
as moft other counties in England, but more particularly
in the hilly or mountainous parts of this province, where they
have the greateft abundance of water.

* This communicated by an old and experienced farmer.

Trials

Trials of throwing water over the lands, have been made in different parts of the county; and it appears, that wherever the trial has been made, and conducted upon proper principles, the attempt has proved highly beneficial to the grounds over which the water * could be thrown, except it had a mixture of metallic, or other noxious matter.

Notwithstanding the fact has been sufficiently proved, in a variety of cases, upon different soils, it is a matter of astonishment, that so rich a source of improvement has been hitherto neglected, when such an extent of ground is capable of receiving the advantage.

The value of water, in this point of view, is not yet sufficiently known †; like many other blessings of life, being, when very liberally bestowed, the less valued. Streams of water, which for ages have passed unnoticed, have within a few years proved a source of wealth to individuals beyond conception. What was probably considered a nuisance, has proved, in many instances, of more real value than the fee-simple of the whole manor, through the vales of which it had so long strayed, by turning machinery, &c.

The many rivers, rivulets, and rills, flowing through the mountainous part of the county, offer their rich streams to meliorate the lands through which they descend. Many thousand acres might partake of their fertilizing effects, at an inconsider-

* See Treatise on watering Meadows, by Mr. Boswell; and Mr. Davis's excellent account of Wilts.

† The value of water is not known in a variety of senses, as it should seem from the following fact: the same freehold had been in possession of the same family for three generations; the present possessor had enjoyed it about fifteen years; and all this while, without having a drop of water for any purpose whatever, but what was carried at great pains for a considerable distance from a stagnant pool upon the head, in a pail. A resolution was however formed, and the work begun in 1794, of sinking a well about two yards distance from the kitchen-door; and the whole work was completed for the sum of seven shillings and sixpence. For this small sum an excellent spring of water has been obtained, and no small portion of labour saved.

In some places, where they are almost drowned in winter, as in Altcar, by the overflow of the river Alt, till lately drained, the families were frequently in such distress as to flee from home, and seek refuge at the Hill-house; and yet, in summer seasons, this country is distressed for want of water, and that to a degree, as to require driving the cattle the space of a mile to drink, the springs being exhausted.

able

able expence; lands too, at prefent poor, barren, and un-productive, at a diftance from other manures, might be rendered competent to maintain an increafed number of valuable animals, by which the quantity of yard-dung would be increafed, and applied, in much more abundant portions, to thofe lands which are beyond the falutary effects of the overflowing waters.

The prefent fyftem of converting the arable into meadow and pafture grounds, to which the water, with peculiar propriety, may be applied, is a ftrong argument in favour of irrigation.

The following neat practice may be worthy of record, as the thought of an ingenious man, game-keeper to R. W. Bootle, Efq. Latham; for which he was honoured with a filver cup, by the Agricultural Society of Manchefter. From the ditches above his houfe, he collects the water, and brings it paft his buildings, from which his lands have a regular defcent. This water carries along with it all the drainings from the farm-yard, which is thrown upon the lands according to the ufual cuftom of irrigating :—but he has funk a refervoir, the fides of which are fecured with pounded clay : in this refervoir he preferves his water, fometimes till a dry feafon ; then throws it upon the land, when the earth wants moifture. He puts marle into the rivulet through which the water runs, and finds it of great fervice.

William Fitzherbert Brockholes, Efq. of Claughton, near Garftang, has made a moft mafterly improvement upon a large morafs, by means of draining and irrigating—it is a good example, and deferves the attention of the farmers in the vicinity : alfo by Mr. Richard Jones, of Peel, in Little Hulton, near Bolton.

CHAPTER

LANCASHIRE BULL.

G. chubbard del.t

CHAPTER XIII.

LIVE STOCK.

SECT. I.—*Cattle* *.

THE Lancashire long-horned cattle are known all over
the kingdom, and found in almoſt every part of the county,
the prime ſtock of which is bred in the Filde, whither the pur-
chaſers from different parts of the kingdom have uſually re-
ſorted; but applications have not of late been ſo frequent as
formerly †. The breed having been almoſt *entirely neglected,*

---

* The following obſervations merit to be preſerved.

To trace the origin of the breed of cattle now prevailing in Lancaſhire, would probably, at this time, be a difficult taſk. But that they
were famous over the whole kingdom, is evident from being ſo frequently
noticed, and in ſuch eſtimation as to be ſought after from all parts of the
kingdom. In ſuch repute were they, and of ſuch ſuperior quality, that
that great judge in cattle, Mr. Bakewell, thought proper to make them the
ſource from which he has, by croſſing, &c. made ſuch improvement. But
as the breed has been under a progreſſive ſtate of melioration in Leiceſter-
ſhire, it ſeems to have been in an equal ſtate of retrogradation in Lancaſhire,
as if over-awed by competition, has ſilently yielded to a conqueror.

It is not long ſince, however, that a celebrated traveller made the following obſervations in his tour through Lancaſhire.

" Breakfaſted at Garſtang, a ſmall town remarkable for the fine cattle
produced in its neighborhood. A gentleman has refuſed thirty guineas
for a three year old cow; has ſold a calf of a month's age for ten guineas,
and bulls for one hundred; and has killed an ox weighing twenty-one
ſcore per quarter, excluſive of hide, entrails, &c. Bulls alſo have been let
out at the rate of thirty guineas the ſeaſon; ſo that well might honeſt Barnaby (*a*) celebrate the cattle of this place, notwithſtanding the misfortune
he met with in one of its great fairs.

> Veni Garſtang, ubi natâ
> Sunt armenta fronte lata.
> Veni Garſtang, ubi malè
> Intrans forum beſtiale,
> Forte vacillando vico
> Huc & illuc cum amico,
> In juvencæ dorſum rui
> Cujus cornu læſus fui."

(*a*) Better known by the name of Drunken Barnaby, who lived the
beginning of the laſt century, and publiſhed his four itineraries in Latin
rhyme.
                              Pennant's Tour in Scotland in 1784.

† Alexander Butler, Eſq. of Kirkland, has frequently ſold young hei-
fers at the advanced price of 50 *l. per* head.

§   the

the pail is become the material object; and as it is an established fact, that animals calculated for speedy fattening are seldom if ever prime milkers, good points of shape and make are less attended to than the milk vein.

Some years ago, the Lancashire breeders suffered those of the more southern counties, as Leicestershire, Warwickshire, &c. to pick and purchase their best stock. Thus the northern breeders lessened the value of their own remainder: and the others made improvements upon that which they had obtained from them on the new principles laid down by Mr. Bakewell, and adopted by Mr. Fowler of Oxfordshire, and others. Nothing valuable is now brought southwardly, out of the more northern counties, once so famous for breeding stock.

Amongst the cow-keepers all varieties are found; they change so frequently, that when a cow, likely to be useful, and at the point of dropping calf, is brought to the market, they purchase it, without paying much regard either to the species or country.

Thomas Eccleston, Esq. of Scarsbrick-hall, has introduced upon his farm the Suffolk polls; and he remarks, they stand the climate, although they have a thin skin and fine coat; and they have, upon proof, been found to answer so well in milking, that frequent applications have been made by the surrounding neighbours to purchase them *.

Mr. Wakefield of Brook farm, near Liverpool, and Mr. George Green of Aughton, have hitherto preferred the Holderness. But the long horn of the true Lancashire breed is the prevailing stock of the county, and seem in general well adapted to the soil; doing less damage to the clay lands, than the heavy Holderness; and being much esteemed by the feeder and butcher for their carcase.

Mr. Orme of Derbyshire, tried nine Holderness cows against nine Derbyshire cows of the improved sort; the former gave the greatest quantity of milk, but that of the latter was considerably more productive of butter and cheese. By the im-

* These stock seem well calculated for the spongy soft lands, being lighter upon the surface than the long-horn.

proved

proved Derbyſhire cow is meant ſuch as was bred by croſſes from Lancaſhire, Warwickſhire, &c. and what the Leiceſter-ſhire breeders and others call the old-faſhioned ſort, before delicacy of fleſh, and the feeding properties, were ſo much attended to. This ſort of cow is generally the home-bred ſtock of Derbyſhire. The milch cows brought by the dealers to Derby market throughout the ſpring in great numbers, are chiefly of the Yorkſhire kind, from the neighbourhood of Rotheram. Theſe the farmer croſſes with a Derbyſhire bull of the above mixed breed. Shortly there will be few bulls in the ſouthern parts of Derbyſhire, without much of the Bake-well, or, which is the ſame thing, the Fowler ſort in them.

More attention is requiſite in Lancaſhire, in the choice of good bulls, than has hitherto been paid by the breeder towards the improvement of his ſtock. Mr. Bakewell has fully con-vinced the world, what may be effected by perſevering atten-tion on this ſubject *.

Of the importance of dairy farming, no doubt can be en-tertained.

It is true that cheeſe may be imported; but milk muſt be raiſed nearly upon the ſpot where it is conſumed, and freſh-butter does not improve by carriage. Milk is the cheapeſt food, and probably the healthieſt, that can at this day be purchaſed. It is no wonder then that the demand for this article ſhould be great in this populous country, and near the great towns on the north-eaſt part.

There is much cheeſe made in this county, and alſo of excellent quality; in many reſpects equal to the Cheſhire, in ſome ſuperior. The cheeſe made in the vicinity of Leigh, Newborough, &c. for its mildneſs and rich flavour, always bears an advanced price at market †; and it is ſomewhat re-markable that the very beſt dairy (as is uſually reckoned) is the very worſt land; the ſoil not being above two or three inches deep. ‡ Superior, if only on the following account—

---

* Mr. Bakewell may have improved ſtock for the *grazier*, particularly where oxen are kept; but who will ſay he has been a friend to the dairy ?

† About 10 *s. per cwt.*

‡ The lands in Weſt Leigh and Weſt Houghton.

U   the

the Lancashire cheese is free from that mixture of colouring matter, which, through the artifice of factors, or the folly of the consumers, particularly those of the metropolis, is, contrary to the inclination and better judgment of the Cheshire dairy-women, infused into the milk. Nay, the factor not only re-fuses to purchase without, but supplies the arnotto at his own expence, which, instead of adding the least benefit, is known to injure the good quality of the cheese: such is at present the in-fatuating folly of fashion.

Many of the Lancashire people, as well as those of other counties, are in the habit of colouring their cheeses, and this is a very growing evil; for this purpose they use foreign ar-notto, but the Cheshire people use English arnotto, which is often made of soap and Venetian red, &c.; the last article is of a pernicious quality.

Dalton, belonging to Richard Wilbraham Bootle, Esq. is unrivalled in Lancashire for cheese, and is undoubtedly the richest tract of land in the county; for, besides being rich fox land, there are infinite beds of stone, flag, slate, and coal. Timber thrives here uncommonly.

Copy of a Letter to the Surveyor, on the Subject of Leigh Cheese.

" The Method of manufacturing Leigh Cheeses; with some Observations on the Quality of the Cheese, the Nature of the Land, and the Quantity made from a Cow.

" I suppose the method of making cheese is pretty well un-derstood, and is nearly the same all England over; but as the cheeses of different countries differ so much in quality, it may be well to enquire from what this difference arises, whether from the method of making it, or from the nature of the land on which it is made; and if both together do not contribute to this material difference.

" The farmers in Leigh parish make their cheese of two

meals

meals of milk, the night's milk and the morning's, fometimes the night's milk is fkimmed, and part of the cream taken from the cheefe, but this not every where, for the beft dairies put all in; in the morning when the cheefe is to be made, the night's milk is to be heated till it is juft as warm as from the cow, and then mixed with the new milk as foon as it is milked;—into this is put a fmall quantity of rennet juft fuf-ficient to come the curd, and no more; for on this juft pro-portioning of rennet and milk, they tell me, the mildnefs of the cheefe greatly depends. The rennet is made from the ftomach-bag of a calf, falted and dried, which they call a bag-fkin; a piece of this, no bigger than a much-worn fixpence, is put into a tea-cup-full of water, with a little falt, about twelve hours before it is wanted, and this is fufficient for 18 gallons of milk, which it will come in about an hour and a half, if the bag-fkin be good; then the curd is broke down, and, when feparated from the whey, is put into a cheefe-vat, and preffed very dry, and after that broken very fmall, by fqueezing it with the hands; the new curd ufed is mixed with about half its quantity of yefterday's, and which has been kept for that purpofe; and a part of this new curd is put by for to-morrow, if it can be fpared; if not, all to-morrow's is put by to mix with new, as convenience fuits, for the beft cheefe is always made with part old curds. Some mix the old and the new together, after both have been worked very fmall: others put the old curds in the middle of the cheefe: either of which ways will do very well, as I have often noticed. When the curds have been thus mixed, and well preffed and clofed with the hands in a cheefe-vat, till they become one folid lump, it is put into a prefs for four or five hours; then taken out of the cheefe-vat and turned, by means of a cloth put into the cheefe-vat for this purpofe, and again put into the prefs, where it ftands till night; then taken out, well falted, and put into the prefs again till morning, when it is taken out, and laid upon a flag, or board, till the falt is quite melted, which will be in a day or two; then it is wiped, put into a dry room upon a turning board, turned every day, till it becomes dry enough for the market. The ufual thicknefs of the cheefe, when dry, is

not

not more than three inches, fo that in five·or fix months it is
hard enough to carry to market; and a great deal of it at this
age is fent to London, by perfons who are commiffioned to
purchafe it from the farmers.   At a year old I think it is in its
greateft perfection, for if it is kept longer it grows too dry;
and for this reafon it is always fold off as foon as poffible it
can be carried without damaging.   The cheefe is mild; and
when toafted it keeps all its butter within it, which makes it
eat foft and rich.   This-property of its mixing together when
hot, is faid to be owing to its being put together cool when
made, for this makes the curd mild and tender, and likewife
the cheefe, fo that its more folid particles, when heated, are eafily
feparated, and the whole fo loofed and broken, that room is made
for the butter, which adheres to the fmall particles of cheefe, and
forms one pulpous confiftence.   Not fo when the cheefe is over-
heated in making, for then more of the butter runs out, and the
curd is fafter bound together than before; and when toafting,
the parts are loofened, the butter is run out, and the remainder
of the cheefe is left hard and dry.—The land round Leigh is
chiefly barren, being ebb of foil and clay under, which makes
it cold and wet.   A few years fince fome of the farmers, encouraged by the high price of corn, marled and ploughed their
farms, which had been grazed time immemorial; the confequence was, the plough foon wore them out, and left them
poorer than ever.   The grafs that came was coarfe and dry, and
the cheefe made off thefe ploughed farms of an inferior quality,
which had like to have brought the whole into difrepute.
But fince the plough was laid by, the paftures have come
about, and the cheefe made upon them begins to fetch as
much at market as the others do.   Of cheefe, the quantity
made from a cow is about 360 lb. fit for the market; befides
a fmall quantity made before and after the proper cheefing
time, which begins when the cows go to grafs, which is generally the old May-day, and ends when they are taken up
for the winter, which is commonly in the beginning of November.

COW.

COW-KEEPING.

The cows kept in the neighbourhood of Liverpool, and within the compaſs of ſix miles, are, after ſupplying the family, principally for the purpoſe of furniſhing the Liverpool market with milk * and butter †. There is milk, it is true, brought to town ‡ from a conſiderable greater diſtance (10 miles) but the general diſtance ſeems no more than what is above ſtated. In the town of Liverpool alone, there are a conſiderable number of cows kept, to the amount of 5 or 600. A ſingle field, for an outlet in the day-time, is procured at a very advanced rate ; but the principal food is hay, and grains from the breweries.—In the town of Mancheſter, at the preſent juncture, there are not more than ſix cows kept within the precincts of the town, for the ſupply of its inhabitants. There comes a quantity every day by the Duke's canal.

Thoſe who are ſuppoſed to follow the beſt ſyſtem of management, with a proper capital, ſeldom keep the ſame cow more than one calf, except ſome particular favourite. They are purchaſed at the time of calving, and the calf is immediately ſold to feeders for the market, and who keep cows for that purpoſe, and diſpoſe of their milk, and procure a livelihood that way. The cows, when they fail of yielding a certain quantity of milk (about 6 quarts per day) are, if in proper

* A few farmers there are that do not carry their milk to market, but diſpoſe of it at home.

† Butter-milk is an article of food throughout the greateſt part of this county. When made into porridge, and thickened with a little oatmeal, and ſweetened with treacle, it becomes an agreeable, nouriſhing, wholeſome, and cheap food : the ſweet, mixed with the acid of the milk, makes it very pleaſant ; mixed with water it is rendered a good beverage at meals, cool, refreſhing, and quenching in ſummer. It is ſometimes mixed with butter, and thus uſed to potatoes.

‡ The conveyance of milk has of late years been in wooden veſſels in carts, inſtead of the backs of horſes, as formerly. One horſe can convey a greater quantity in a cart, with more eaſe, than on his back, beſides affording more comfortable accommodations to the good woman, who alſo can carry along with her milk ſome little garden-ſtuff, according to the ſeaſon of the year ; and there are but few milk-carriers that do not take a few greens, &c. from their gardens, which they can diſpoſe of amongſt their cuſtomers, whilſt they are ſelling off their milk. Of late theſe milk-carts have been covered with painted canvas upon hoops, affording a very good ſcreen from the ſeverity of the weather.

condition,

condition, difpofed of to the butcher; and, if properly kept, to advantage, i. e. for more than the firſt coſt. Mr. Mayo, who has a milk farm upon the eſtate of T. Butterworth Bayley, Efq. of Hope, near Mancheſter, informed us, that he generally fold his ſtock off to the butcher, at an advance of two guineas more than the original purchaſe *. But his landlord has furniſhed him with the greateſt conveniencies, and the completeſt farm-yard obſerved in this furvey †, from which he has profited, and merits praiſe for his great induſtry and excellent fyſtem, which is to feed them with the choiceſt hay, and opening food in winter; tempting their appetites, by offering his cows only ſmall quantities at once, but this is frequently repeated; and during the ſeaſon they are upon grafs, they eat corn, ground or bruiſed in a mill, of ſuch different kinds as can beſt be purchaſed; a very ſmall portion of time is employed in grazing, for being well ſupplied in the ſtalls, and from the luxuriant rich grafs in the fields, they lie at their eafe and ruminate. Mayo generally keeps his cows about 18 months, and contrives to fell off in the ſpring, when beef is at the deareſt.

* " I do not undeiſtand the mode of Mayo's cow-keeping, to keep his cows only one note of milking, and generally to keep them 18 months; which way he keeps them to profit is a myſtery to me that I cannot find out, and felling them at two guineas per head more than the firſt coſt.
" The general mode with me, and the cowkeepers in and about Liverpool, when they are kept only one milking note, is from ſix to nine months at the longeſt, for that is as long as any cow can pay at one note, except a prime cow that may be kept for ſeveral calves, or as long as ſhe does well. Cows, that only milk one note and calve, from July to November and December, are moſtly turned off in the ſpring following, when beef begins to be more ſcarce, and to fell at the deareſt; by which means the paſture is eaſed at the beginning, when milk is the moſt plentiful, and coming in great plenty out of the country at a far diſtance. Some cows that are turned off, when beef is felling the deareſt, and with additional keep, may fell for more by two or ſometimes three guineas than the firſt coſt; and as that additional keep is feldom or never accounted for, it always appears as if it was ſo much clear profit; but without the debtor and creditor account be clearly ſhewn, it is not fairly explained to the public. I generally fee grofs amounts given of prime cows, but never the amount of an improving cow, of which there are more than prime cows, particularly for milking."—Mr. Harper.

† Among other conveniencies, a ſtream of fine water runs through the yard; and by opening a cock, he can throw a ſtream through the cowhouſe, to waſh away the dung, &c. left after emptying, and this water is obtained by draining the higher lying lands.

All

LANCASHIRE COW.

All the cow-keepers do not follow this good practice; and some, who regularly change their cows, do this frequently at the loss of two guineas *per* head.   A cow at dropping calf, is generally worth, *cæteris paribus,* two guineas more to the cow-keeper than she would be to the butcher.—If she can be sold after nine months milk, for the first cost, or any advance, it must depend upon the beast being well bought, the season of the year when sold, or extra keep to promote feed.

1794. October 22d.   Mr. Edward Ashcroft, farmer, at the Spellow house in Walton, sent down, for the surveyor's inspection, the butter of a prime cow, collected the preceding week, the milk kept by itself, and churned the day before: the amount of which was 16 lb. of butter of 18 oz each.   The butter had a fine yellow colour, and acknowledged by all who viewed this great curiosity an excellent specimen.

The cow which yielded this astonishing quantity of butter, has had five calves, is eight years old, calculated to weigh about five hundred.   The colour a light red, a good deal of white, of the Lancashire breed, a very long horn, which was un-usually thick towards the root.   She had calved about a month before; her food eddish, but not of the first bite, or best quality, with grains from the brewery, or scalded bran; the quantity of milk she gives at present *per* day 22 quarts; a specimen of which accompanied the butter, and was tried by Dicas's lac-tometer, and which was 96; after standing 30 hours, and the cream taken off, 103.   There seems no superior richness in the milk, therefore the great quantity of butter arises from the large quantity of milk yielded.—But this is her prime season, she will gradually fall off in quantity, but not it is said infe-rior to the general quantity by the remaining stock.

It must appear astonishing that notwithstanding the progeni-tors of this beast possessed, and her successors still inherit, the good qualities of this prime cow, yet there appears an indolent negligence in the propagation of the breed.   A bull has never yet been thought of to propagate from full blood; for, besides this disposition to milk, when that can be got rid of, there is a general disposition to fatten.

The Liverpool cowkeeper does not aim at making butter;

9   his

his fyftem is, to fell milk and cream; but in the fummer feafon, when milk flows into the town from many quarters, a
market fufficient to take off the whole may not always be
found, and then he is under the neceffity of churning it, and
making butter, or difpofing of it in cheefe, or fome other way;
but the confumption of milk and cream is univerfal; and to
thefe two articles his greateft attention is directed.

A good cow fhould give daily 12 quarts, and the price of
cream is generally 14 *d. per* quart; new milk 2 *d. per* quart,
and inferior milk 1 *d. per* quart *. A cow ftands the keeper
in about 1 *s.* per day, for food, atttention. &c. fo that with
contingencies, and loffes that frequently happen to the ftock,
there is but barely a living profit † left to a bufinefs, which
requires much attention, and not a little fkill in purchafe and
management.

> Mr. Henry Harper's ftatement of the expence of keeping,
> and produce of a cow per ann. averaged out of a ftock
> of twenty-five cows, kept upon the Bank Hall eftate.
> —The fales of produce, and expence of keep, according
> to the prefent price of the different articles mentioned,
> 1794.

	£.	s.	d.
Average butter of one cow for 52 weeks is 4 lb. per week; 208 lb. of butter, at 11 *d.* per lb. - -	9	10	8
Milk of all kinds, 52 weeks, at 3 *s.* 3 *d.*	8	9	0
Price of calf - - - -	0	4	0
Three tons manure, at 4 *s* 6 *d.* -	0	13	6
Cartage faved, by the dung on the premiffes. - - - -	0	7	6
	19	4	2

* Dearer at Manchefter market a trifle; probably the quality may be fuperior.

† In calculations we too frequently find that no allowance is made for contingencies, or falling off of quantity. Twelve quarts *per* day is the prime milking quantity; and though fome cows may have given more at the firft, thefe kind of ftock more rapidly fall off in quantity, whilft, at the fame time, the quality was of lefs value, in proportion to the excefs of quantity.

Expences

	£.	s.	d.
Expence of grafs for the fummer -	2	5	0
Hay, 160 ftone, at 8 ½ *d.* - - -	5	13	4
* Provender, 26 weeks, at 3 s. 6 *d.* -	4	11	0
After grafs or eddifh - - -	1	10	0
Lofs in cattle 5 per cent. 9 s. per head	0	9	0
Cart-horfe and keep (to carry the produce to market) - - -	0	2	6
Dairy-maid - - - -	1	0	0
Attendance to milk - - -	1	2	6
Wear and tear, mugs, &c. - -	0	1	6
Salt for 208 lb. butter, 16 lb. -	0	1	9

		£.	s.	d.
		16	16	7
Profit per ann. -		2	7	7

The average milk of Mr. Harper's ftock is feven quarts of milk per day the year through; although fome prime cows in their full perfection, and in the height of grafs, may yield when frefh calved eighteen, twenty-four, or even thirty quarts, of milk in a day; but this fuperabundance is but of fhort duration.—From every twelve quarts of milk is produced one pound of butter, of 18 oz. to the lb.

* The provender confifts of two feeds, morning and evening, each day, half a peck of potatoes or turnips cut and given raw, value one penny halfpenny; one pint of oats and one pint of barley mixed together, and boiled with chaff, cut ftraw, bran, or malt-duft, mixed with the potatoe or turnip, value one penny halfpenny, or three pence each meal. The corn is boiled in plenty of water till it burfts, and the water is ufed in the mixture.

X

Average

Average Price of BUTTER in the Liverpool market from the same, for the years 1791, 1792, and 1793.

1791.			1792.			1793.		
Weeks. d.	s.	d.	Weeks. d.	s.	d.	Weeks. d.	s.	d.
7. 12 —	7	0	16. 11 —	14	8	18. 12 —	18	0
19. 11 —	17	5	4. 12 —	4	0	8. 13 —	8	8
2. 11½ —	1	11	12. 10 —	10	0	11. 11 —	10	1
1. 10½ —	0	10½	1. 13 —	1	1	4. 10½ —	3	6
18. 10 —	15	0	9. 11½ —	8	7	7. 11¼ —	6	8½
5. 9 —	3	9	8. 9½ —	6	4	2. 14 —	2	4
			2. 10¼ —	1	9	2. 12¼ —	2	1
52 Weeks -	45	11½	52 Weeks -	46	5½	52 Weeks -	51	4½

Average in 1791.	Average in 1792.	Average in 1793.
10½ $\frac{12}{52}$ per lb.	10¼ $\frac{42}{32}$	11¼ $\frac{32}{52}$

Average price of butter for the three years is, per lb. 11 $\frac{6}{13}$ d.

The fyftem at Manchefter is nearly the fame as at Liverpool, (fee the preceding note upon Mayo's good management). It does not, however, appear, that fo many cows are kept within that town, it being fupplied by a whole circle of furrounding country; whereas Liverpool has only half the quantity of land, from its maritime fituation. The price of labour too, about Manchefter, is fuch, that the milk paffes through the hands of retailers, who buy it wholefale from the farmers,—who carry it generally upon horfes, and whofe fervant, upon difcharging his load, can immediately return and become ufeful at home.

There have been lately introduced milk cifterns, formed out of a black clofe-grained ftone, fomewhat fimilar to black marble. The Rev. Mr. Johnfon has one, containing 13½ feet, another 15½, the expence of which were 2s. 2d. per foot, thefe are furnifhed with lead pipes at that price. Thefe cifterns are remarkably neat, and eafily cleaned.

7

The

The practice of managing the milk for butter in this county, might be of service, if followed in other places. Except in the county of Chester, it should seem (as the surveyor understands) peculiar to this district. The mode is, dividing the milk into two parts; the first drawn, being set apart for family use, after being skimmed; the cream of which goes into vessels appropriated to receive it; as also the whole of the second, or last, drawn milk, provincially called *afterings**; these two being mixed together, are stirred, but not a great depth, to prevent the bad effects of foul air accumulating on the surface: and kept, according to the season of the year, exposed to the fire, to bring on fermentation and sourness; which is accelerated by that which may remain in the pores of the vessels; to prepare this fermentation, they are not scalded, except after having contracted some taint: and then to accelerate it (the quicker it is the better) the vessels are sometimes rinsed out with sour butter milk; in which state the milk is ready for the churn; and, in consequence of this treatment, more butter is obtained, and of a better quality, than if the milk was churned sweet. And the butter-milk, as it is called, after the butter is extracted, instead of being given to the hogs, as is generally the practice in many counties, becomes, under this process, an excellent food for man, both wholesome, and pleasant, as before-mentioned. This is the sort of butter-milk which, it has been remarked, is necessary for such labouring poor as live on potatoes.

EXPERIMENTS REGARDING MILK.

Thomas Wakefield, Esq. Brook-farm †, about two miles

---

* About one half from each cow, each meal; but the quantity taken first in some measure depends upon the consumption of milk in the family.

† Mr. Wakefield has applied the steam of warm water for some time past, in his stoves; and, by its effects has produced some very luxuriant fruit, both pines and melons. Mr. Wakefield seems to think that a new field, in the process of vegetation, may be discovered through the means of this application. But he is preparing to lay before the public the particulars of the process, and its effects.

and

and a half from Liverpool, keeps a regular account of the produce of his milk, butter, and amount of fales, pofted up every
fortnight; with remarks upon the effects of different food,
change of weather, or any other particular caufe, which may
occafion any confiderable variation in the amount of the different produce.   Thefe remarks are entered into the margin —
from thefe regifters the furveyor has been favoured with the
following extracts:

1ft. An experiment made on feven cows, for three fucceffive weeks.   Firft week, they produced 289 quarts of milk.
This week he took only one pint of drippings, or afterings,
from each cow, each meal; which, together with the cream of
the former or fore-milk, produced 25¼ lb. of butter.

The amount of this week's fales of fweet and churned milk
and butter, from this method, was £. 2. 7 s. 4 d.

2d. Second week.   The fame cows produced 294 quarts of
milk.   This week he took half of the milk each cow gave
each meal, as afterings or drippings; thefe, with the cream of
the fore-milk, produced 28½ lb. of butter.

Amount of fales this week, from this management, was
£. 2. 4 s. 2 d.

N. B.—Although there was more butter produced, there
was lefs new milk brought to market.

3d. Third week.   The fame cows produced 287 quarts of
milk.   This week he took only half a pint of drippings from
each cow each meal, which, with the cream of the fore-milk,
produced 23¼ lb. of butter.

Amount of fales this week was £. 2. 5 s. 4 d.

N. B.—The fore-milk, or firft-drawn milk, is put into leaden
cifterns, and is found to anfwer beft, if not above three inches
deep.   The amount of fales includes the amount of fweet-
milk, butter-milk, and butter, as produced from new-milk.

From the foregoing experiment it appears, that though the
fecond week's produce of both milk and butter was the greateft, yet the amount of fales was the leaft; which deficiency
arifes from the fmall quantity of fkim milk, by churning fo
much afterings.   Butter-milk being only ½ d. per quart, fkim-
milk 1 d.

4th.

4th. From the 1st of May 1790, to 30th April 1791, 100 cows produced 271,270 quarts of new milk, 23,632 lbs. of butter, and amount of sales £. 2,854. 2 *s*. 9 *d*. . .

It would have been satisfactory if the foregoing curious statements, had been attended with a regular debtor and creditor account, with profit and loss, account of sales of cattle, with a number of other particulars; so as to have clearly stated the clear gains of such large gross receipts.

5th. The following statements may prove the advantage of regular churning, or rather disadvantage of irregular work. These operations being so very heavy, it became too much for a couple of men to support, which occasioned a machine to be procured, a cog-wheel, &c. and by which is effected, with a horse and a boy to drive, in one hour and a quarter, what was usually the labour of two men five hours .

Quantity of new milk.	Quantity of butter by hand-churning.			
Quarts.	Pounds.	£.	*s*.	*d*.
6,471 - -	364 - -	47	1	7 } Amount
6,644 - -	397 - -	49	0	9 } of
6,995 - -	348 - -	49	0	9 } Sales.
Quarts 20,110	Pounds 1,109	£. 145	3	1

Quantity of new milk.	Quantity of butter by machinery.			
Quarts.	Pounds.	£	*s*.	*d*.
7,261 - -	469 - -	55	4	1 } Amount
7,675 - -	482 - -	56	14	11 } of
8,120 - -	574 - -	65	3	8 } Sales.
Quarts 23,116	Pounds 1,525	£. 177	2	8

The above quantities of milk were the produce of six successive fortnights.

* *Hand Churning and Machinery.*

" There can be no difference in the churning, if the hand-churning be worked brisk till it offers for butter f prepared in the same manner, which always may not have been the case with Mr. Wakefield, therefore machinery may have the advantage with others as well as Mr. W."—*Mr. Harper.*

There-

Therefore if 20,110 quarts yield 1,109 pounds of butter, how many pounds will 23,156 quarts yield?

Anſwer 1,277

1,525 produced by machinery

248 pounds more than would have been produced by hand-churning; which, at 10 *d.* per lb. is £. 10. 6 *s.* 8 *d.*

Quarts.	£.	*s.*	*d.*	Quarts.
Again, if 20,110 ſell for 145	145	3	1	what will 23,156 ſell for?
Anſwer	167	2	8	
	177	2	8	did ſell for.

£. 10   0   0 profit by new mode of churning.

Again, if 23,156, gain £. 10. what will 271,270 quarts gain?
Anſwer   -   £. 117.   2 *s.*   11 *d.*

Hence it appears, that a churning machine, on one hundred cows, in twelve months, will gain £. 117, beſides the expence of labour.

A ſhort-horned cow, upon an average of twelve months, yields nine quarts of milk per day, and 4½ lb. of butter per week.

A Lancaſhire long-horn yields eight quarts of new milk per day, and four pounds of butter per week for twelve months.

N. B.—In making the foregoing experiments, the cattle have had always the ſame kind of food. But to know the clear reſult, the quantity of food conſumed by the two breeds of cattle ſhould be clearly aſcertained, before any deciſive concluſion can be drawn. The produce of milk and butter is in favour of the Holderneſs—neat balance, not yet apparent, whether in favour of long or ſhort-horn. The fleſh of the latter is ſaid to be of inferior quality.

THE SURVEYOR'S EXPERIMENTS.

I directed the uſual quantity of milk generally churned at one ime, and collected according to cuſtom, to be meaſured previous

vious to the operation: 15½ gallons milk, three pints warm water added.   After the butter was extracted, the milk meafured again thirteen gallons five pints.   Quantity of butter produced 8 lb. 4 oz.

Again, directed the cream from all the cows for the fame fpace of time only, to be collected and churned without any other milk.   Quantity, cream four gallons, and three pints of water added.   Produce of butter, 4 lb. 14 oz.; of milk, after butter was extracted, four gallons one pint.

Obfervation.   More butter, from quantity, in the laft experiment; but a great deficiency of butter this week from this mode.

Lefs quantity is loft by extraction of butter than might have been expected, confidering abforption of veffels, fplafhing over of milk, &c.

Both thefe experiments prove the great advantage of felling cream at 14 *d.* per quart, in preference to churning.

	£.	*s.*	*d.*
Ergo. Firft, fay butter worth, at 12 *d.* per lb. -	0	8	3
Butter-milk, at 2 *d.* per gallon, worth - -	0	2	2¼
	0	10	5¼
But the milk of the firft 62 quarts, even at 2 *d.* per quart only, without the trouble of churning, was worth - - - - -	0	10	4
Again, 4 lb. 14 oz. butter worth, fay - -	0	4	10
Butter-milk 4 gallons 1 pint, at 2 *d.* per gallon -	0	0	8¼
	0	5	6
But 4 gallons of cream, at 4 *s.* 8 *d.* per gallon. or 14 *d.* per quart, worth - - -	0	18	8
In favour of cream - £.	0	13	2

Upon his farm at Aughton Mr. G. Green obferves, that the average milk by his cows has been nine quarts of milk by the fhort-horn, and feven quarts of milk per day by the long-horn cows; and of butter eight pounds per week by the former and

feven

feven pounds per week by the latter. This quantity is three pounds per week more than either Mr. Wakefield's or Mr. Harper's cows yield, which are equal in quantity, namely each 4lb per week. The two farms are about equal diftances from Liverpool, *e. g.* Bank Hall, two miles north weft.—Brook Farm two miles fouth eaft.

LACTOMETER.

A lactometer, to try the different qualities of milk, has been invented by Mr. Dicas, mathematical-inftrument-maker, in Liverpool, and patentee of a neat, fimple, and acurate inftrument to try the ftrength of fpirituous liquors and worts.

This lactometer afcertains the richnefs of milk, from its fpecific gravity, compared with water, by its degree of warmth taken by a ftandard thermometer, on comparing ts fpecific quality with its warmth : on a fcale conftructed for this particular purpofe, and by which, if the principle be right, may be difcovered not only the qualities of the milk of different cows, paftures, foods, as turnips, potatoes, grains, &c. but alfo probably which may be the beft milk, or beft paftures for butter, and which for cheefe. This inftrument, however, is yet in its infancy. The furveyor took one with him upon his journies, and made experiments at different places; but time fufficient for a full and complete experiment feldom offered : other circumftances intervened, and prevented a fair trial; but, at his own houfe, he has made a number of varied experiments, upon different milks from different farms.

Obfervations on the conftruction of the LACTOMETER *for determining the goodnefs of Milk, and the advantages to be derived therefrom: By Mr.* DICAS.

The LACTOMETER is conftructed with ten divifions upon the ftem (fimilar to the patent brewing-hydrometer,) and with eight weights, which are to be applied only one at a time upon the top, to obtain the weight of the milk : an ivory fliding-rule accompanies this inftrument, upon the middle or fliding part of which is laid down the lactometer weight of the milk, going

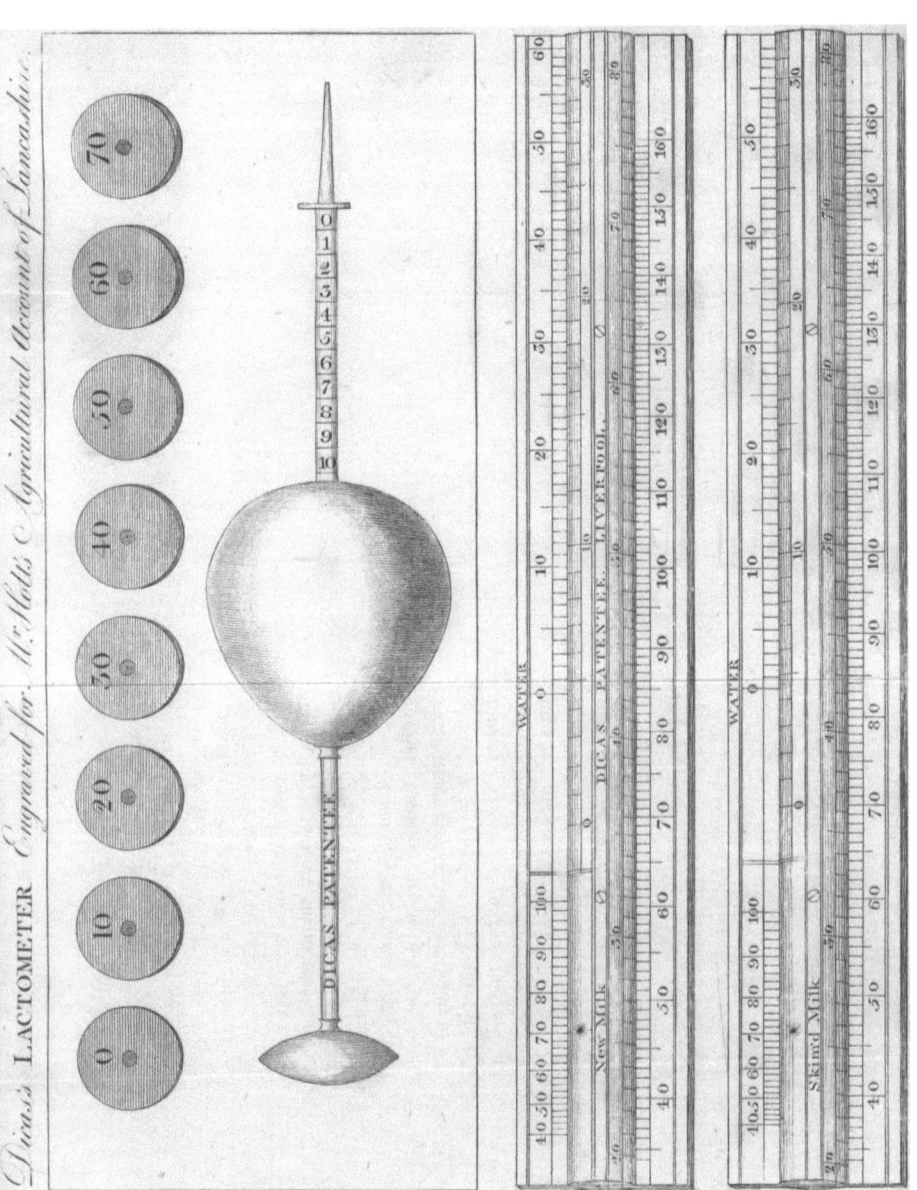

Dicas's LACTOMETER Engraved for Mr Holt's Agricultural Account of Lancashire

The material originally positioned here is too large for reproduction in this reissue. A PDF can be downloaded from the web address given on page iv of this book, by clicking on 'Resources Available'.

from 0 to 80; and oppofite thereto are placed the various ftrengths of milk, from water to 160—100 having previoufly been fixed upon, from a number of experiments, as the ftandard of good new milk, and each of the other numbers bearing a proportionate reference thereto.

At one end of the fliding-rule the degrees of heat from 40 to 100 are placed with a ftar oppofite, as an index to fix the flide to the temperature of the milk.

The whole being graduated to fhew the exact ftrength of the milk as it would appear in temperature of 55 degrees of heat, although tried in any inferior or fuperior temperature between 40° and 100°; thus the great inconvenience which would attend bringing the milk at all times to one temperature is avoided, and a fimple mechanical method of allowing for the contraction and expanfion fubftituted.

And as fkimmed milk, being divefted of the particles of butter which exifted before fkimming, appears to have a lefs degree of affinity with that than the new milk has, one fide of the ivory fliding-rule is adapted to fkimmed, and the other to new milk.

GENERAL RULE.

Firft, find the temperature of the milk with the thermometer, and fix the fliding-rule fo that the ftar fhall be facing the degree of heat the mercury rifes or falls to; then put in the lactometer and try which of the weights, applied to the top, will fink it to fome one divifion upon the ftem; add the number of the weight upon the top, and that of the divifion together, and oppofite the fame formed upon the fide, will be fhewn the ftrength of the milk.

EXAMPLES.

OF NEW MILK.—If in the temperature of 72°, the lactometer with the weight 40 finks to 9 upon the ftem, fix the flides fo that the ftar fhall be facing 72°; then oppofite 49 will be found 100, the ftrength of the milk.—Again, if in 60° the lactometer with 50 on the top, finks to 6 upon the ftem, the flide being fixed for new milk fo, that the ftar fhall be at 60° of heat, then

Y   facing

facing 56 will be found 110, the ftrength of this milk in proportion towards the other, provided it is equally replete with
cream.

To difcover which, it becomes requifite thefe two famples
fhould ftand a certain time that the cream may rife, which being taken off, they are to be tried with the lactometer again;
and as the cream is evidently the lighter part, the milk will appear by the lactometer denfer or better in quality than before.
Suppofe the milk in the firft example to be 57 by the lactometer in 60 degrees of heat, then the ftrength by the fkimmed-
milk fide of the rule will be 112. And admit the fecond example of new milk to be 58 in 64° when fkimmed, the ftrength
would be 116.

As a comparifon—

Say No. 1. New milk	-	-	-	100
Ditto fkimmed		-	-	112
		Difference	-	12

No 2. New milk	-	-	-	110
When fkimmed	-	-	-	116
		Difference	-	6

From which it appears that No. 1 has produced a larger
quantity of cream than No. 2, and confequently may be deemed
the better milk.

Some inftances have occurred where the ftrength of new
milk has only been about 80, and when fkimmed near 100.

Thus it may, without the leaft impropriety, be called a milk
much better adapted for making butter than cheefe. And the
experiment No. 2, a milk more advantageous for cheefe than
butter, it being confiderably denfer, and confequently containing a much larger portion of the curd, or more folid parts,
which conftitute the bafis of cheefe. The ferum or whey in general being near the fame denfity.

Inftances wherein the LACTOMETER *may be ufeful.*

In difcovering what breed of cattle are moft advantageous.

What food in the winter feafon, whether carrots, turnips, potatoes, &c. are beft.

What the effects of different paftures may be.

How far particular farms are beft adapted to making butter or cheefe.

How far the inconvenience of large cheefes in fome dairies being too rich to ftand, may be prevented, by difcovering when this redundancy of richnefs exifts in the milk.

And in fixing a ftandard for the fale of this ufeful article of life.

A ftandard for fkimmed milk may readily be fixed by faying what ftrength the common faleable fkimmed milk fhall be by the lactometer, or what its fpecific gravity fhall be in relation to that of water in the temperate degree of heat, and that an eafy comparifon may be made between the fpecific gravity of any milk, and its lactometer ftrength: this inftrument is fo conftructed that one of fpecific gravity fhall exactly correfpond with three of ftrength—that is, the ftrength of 90 by the lactometer is a milk whofe fpecific gravity is 1030, to common pump-water 1000.

From a number of experiments and obfervations, the common faleable fkimmed milk in Liverpool is from 52 to 64 of ftrength, and that of new milk from 70 to 80; but it would be difficult to fix any ftandard for the latter, unlefs fome mode could be devifed to difcover whether it was mixed with old milk or not. The only method would be, after fixing the ftrength of it, to try, by letting it ftand, to difcover if it produced that quantity of cream, which, as new milk, it might reafonably be expected to do.

FEEDING

FEEDING CATTLE.

The following practice, by an experienced farmer, (Mr. Henry Harper, Bank Hall) is given in his own words.

" I HAD one year six cows that I house-fed, all at one time, and nearly all of an age; and by way of experiment, I fed two with turnips and ground corn; and two with boiled potatoes and ground corn; and two with raw potatoes and boiled corn: they were all put to feed at one time, and when I thought them fit for the market, I fold three, one from every lot, and went to fee them dreſſed. In thoſe two fed with ground corn and turnips, and ground corn and boiled potatoes, there was little or no difference; but that which was fed with raw potatoes and boiled corn, was better in fleſh, and more fat within fide, than the other two, by a fortnight's keep; and this was not only my opinion, but the butcher's who killed them: the other three I kept three weeks longer; and when killed, they were proportionably nearly in the fame ſtate with the others, but better by being kept the longer; ſo I prefer boiled corn to any ſort of grain, and think it more forcing, either for milk or feeding. They had all one and the fame quantity of corn, &c."

Boiling corn has been practiſed by ſome others, with good fuccefs. A little linſeed improves the quality. Hay feeds, that drop out of the hay, fhould be carefully preſerved, and worked up in mixtures of potatoes or oats, either ſcalded or boiled. The furveyor has experienced the good effects of hay-feeds upon his cattle, for many years; an ingenious farmer, lately talking upon this ſubject, obſerved, that the feeds of many weeds might be converted to good uſe; and ſpoke with confidence of the feeding quality of ſome of them.

Inſtead of oil cake, the lint feed boiled, and inſtead of ſpent grains from the breweries, barley boiled and mixed together, with the addition of chopped ſtraw, hay-feeds or chaff, have been tried by Mr. J. Balmer, of Toxteth Park, both upon milch and feeding cattle; and with more profit than with either of the reſiduums.

Method

Method of feeding Cows, by MR. HENRY HARPER.

There are feafons in which it is fo very difficult to make good hay, that much will be damaged although the greateft attention be paid. The confequence of which is, the milk given by the fame cows is lefs in quantity, and of inferior quality; the butter both lofes its natural colour and good flavour; to remedy which, Mr. Harper takes the following method.

He provides fome fort of provender for his cows; that is, fome fpecies of ground grain; and to mix with it, he procures fome hay of the beft quality, and from the moft fertile lands, which he treats in the following manner. This rich hay is to be ufed as an ingredient for tea *, by pouring boiling water upon it; and the infufion he makes ufe of to fcald his ground grain, chopping the hay, before it is infufed, with an engine, defigned for the purpofe of cutting ftraw ; and this hay, fo cut to the fize of one inch long, is to be mixed with fcalded provender, to the amount of two or three quarts to every beaft. This mixture of bruifed grain, fcalded with the infufion of rich hay, and the addition of the hay to the amount of two or three quarts to each beaft, improves the flavour of the butter, and reftores it to its proper yellow colour.

The milk cows in general, not in the vicinity of towns, are wintered moftly upon hay. Were they, according to circum-ftances, fed with turnips or cabbages, they would be kept at lefs expence to the farmer, and fummer fallow be exploded. Some few, who have begun to fow turnips, fell the overplus to their farming neighbours at from 6 *d.* to 8 *d.* per bufhel, which has produced from thirty to 40 *l.* and upwards, per acre, eight yards to the perch.

---

* " If hay be damaged, it is not proper food for milk cows ; and making good hay into tea is both tedious and unneceffary, as the ftomach of the cow will beft digeft the food, and do all that is neceffary ; and in my opi-nion, the beft engine for chopping hay is in the cow's mouth, which nature has provided. True it is, the better a cow is kept, the more milk and butter fhe will give. If damaged hay cannot fafely be given to the young cattle, it may be ufed as litter."—*J. B.*

Vegetables

Vegetables boiled for Cattle.

Before concluding this article, it may be proper to obferve, that a college of Roman catholics refiding at Stony Hurft, near Clithero, in this county, keep their horned cattle within doors, and fed them upon boiled vegetables; amongft which were included all forts of weeds, dock, nettles, &c. It is well known that on many parts of the continent they feed their cattle on the leaves of trees.—What a refource here opens for the attentive and fkilful agriculturift!

Sect. II.—*Sheep.*

This is not a fheep diftrict, therefore they cannot be any where numerous in the county.—There are flocks (but flock is an undeterminate number) it is true of half-ftarved crea-tures upon the mountains, but in fuch proportion, that Mr. Ecclefton is of opinion there is not a fingle fhepherd, properly fo called, in the whole county.

Thofe which are kept upon the feeding diftricts are bred in Scotland, and purchafed by the Weftmorland farmer from thence at a year old, and afterwards by the Lancafhire grazier from Weftmorland at four years old, fatted and fold for flaugh-tering.

There is a fingular cuftom prevails in the northern part of the county, and which is univerfal amongft the mountains and wafte lands, which is as follows: Whenever a tenant enters upon a farm upon which *there is a heavy-bred flock of fheep,* that the fheep are feparated and forted; viz. the wethers aged, ewes, one year old (provincially hogs) two years old (twinters) and then valued at certain but different prices; and the te-nant by covenant in his leafe to leave an equal number of each fort upon his farm when he quits, or to pay the value in money, according to the deficiency which may appear in each fort; but if proved, on ftating a balance, that it is in favour of the te-nant; he either paid for the overplus number, or his landlord takes them at a proper valuation.

I   The

The sheep are generally delivered to the coming-on tenant about Martinmas, and marked when delivered with red (a species of ochre) in the forehead. The red is provincially called *smit*; and every different farmer marks his sheep upon the back, buttock, shoulder, or in some other part, in a different manner from his neighbour, and also cuts the ear of his sheep, when lambs, different from the other, as a mark of distinction between the two flocks; these two marks, that upon the ear, and the other upon the wool, are never altered, that is, each farm preserves its own peculiar mark, lthough the tenant be changed, and is looked upon as hereditary to the estate. Initials of the owner's name are avoided, though fonetimes practifed, becaufe the largenefs of the mark depreciates the value of the wool.

In the mountainous parts of the country, fleece wool, weighing 16 *lb.* the ftone, fells for 7 *s.* Skin wool at 8 *s.*

" Sheep delivered to a farmer, when he enters upon a farm, are valued at about 8 *s.* When fold to the butcher, from the common, 10 *s.* 6 *d.* and when fatted in the inclofed ground from 16 *s.* to 21 *s.* As to the quantity kept on commons, it is very hard to afcertain, becaufe there is fo much difference between the high commons and the low ; for inftance, on the high commons, fuch as Seathwaite fell, not more than four or five upon an acre; inclofed land in Lowfurnefs, is allowed to fat feven or eight on an acre, but thefe are twice the weight of fell sheep. Thefe are frequently fold from 32 *s.* to 40 *s.* per sheep.

There are but few sheep kept in the southern part of the county, except thofe purchafed in diftant parts, by the butchers, and kept a few weeks on grafs for their own convenience—or, by a few gentlemen †, for the convenience of their families, curiofity, or occafionally to feed upon, or eat off, their turnips, previous to laying down the land. In the northern part of the county, sheep are bred and kept upon the

---

* Generally of Culley's breed from Northumberland.

† Mr. Ecclefton, before mentioned, has a Spanifh ram, a prefent from his Majefty, which has already improved the quality of his wool.

mountains

mountains and moorland. There is also a breed called the Warton, or Silver-dale cragg sheep, which is much esteemed for the fine flavour of its flesh, fineness of its wool, and tendency to fatten. They pasture upon very rocky lime-stone land. Their wool commonly sells at about twelve shillings per stone, of 14 lb.

Annotations.

1. The small number chiefly owing to the number of *dogs* kept in the towns, and universally by the cottagers in the manufactories.

2. The practice of plashing hedges is almost unknown in the county *, and the fences are in general so wretchedly bad, as to render it impossible for the farmer to keep sheep, for which stock a great part of the county seems calculated. Another great objection to the encrease of sheep in this populous county is owing to the great number of dogs, which frequently do great damage to the flocks; but which a tax upon dogs might prevent.

3. Sheep would answer extremely well in many parts of this county; but the Lancashire people are perfectly ignorant of there existing any other species than the black-faced Scotch and Welch sheep: animals active enough to clear a six-foot wall, consequently that cannot be restrained by such infamous fences as are prevalent throughout the county. The application of sheep to turnips, is considered as a caprice that may suit the pocket of a gentleman, but inconsistent with the finances of a farmer.

4. The lands in general, in the southern parts, are extremely proper for sheep, and produce most excellent crops of turnips; but they are not much sown, owing greatly to the common clauses in the leases, of not allowing clover stubbles to be sown with wheat, for which the soil seems very proper.

* Mr. Eccleston's plashed fences are specimens of great neatness and attention to that business.

J. Chubbard delt.

LANCASHIRE MARE.

SECT. 3. *Horses.*

A GREAT number of horfes have of late years been bred, owing to the advanced price they have generally fetched at market; but proper attention in the choice of either the brood mares, or ftallions, has not been paid. The ftocks, both of cows, of fheep, and horfes, are capable of great improvements, which merit the confideration of every breeder.—The fame pafture will rear the young ftock, of either cow, fheep, or horfe, of the beft kind, at the fame coft as a ftock of the very worft quality; but a three-years-old heifer, of the firft kind, will fell for double the price of·one of a fimilar age of the latter defcription; if a colt, the proportion is ftill higher, according to the fuperiority of its breed. If the above ftatement be true, is it not to be wondered at, that greater attention has not been paid by the breeder; fince both the climate and lands are capable of producing good breeds, and there are purchafers enough to excite encouragement? Strong horfes are moft in ufe, except among gentlemen, who breed for themfelves.

Horfe-furgery of late, under Mr. Moorcroft, and by the eftablifhment of the Veterinary College, feems making rapid progrefs towards a degree of perfection unknown in other countries.

Unfortunately no attention has been paid to the difeafes of neat cattle, fheep, fwine, &c. Were the noftrums of individuals for thofe animals communicated to the Board, probably there would be found fufficient remedies for the diforders they are liable to.

Should another circular letter ever be emitted by the Board, might not that be a proper article for enquiry? Or would it not be advifeable to fend a circular letter to practitioners in the farriery line, and farmers, &c. &c. fpecifying each diforder; and by way of encouragement, to grant honorary rewards or medals to fuch may make known the moft fatisfactory receipts for cure or prevention?

Mr. Ecclefton fuggefts the following hint. He imagines, that the number of horfes bred in this, furpaffes that of any

Z   other

other county in the kingdom *.   He propofes, " that a yearly
tax be laid upon ftallions of five times the fum † they receive
for ferving each mare, for the feafon; it would prevent the ufe
of the inferior fort of ftallions, which only ferve to procreate
thofe of fmall value which are nearly ufelefs, with which almoft every part of the kingdom abounds.   A very confiderable
fum would annually accrue to government, were each ftallion
to pay five times the fum for a licence, that he ferves each mare
at, viz. a horfe that covers at one guinea for the feafon, fhould
pay five guineas for a licence; and others, that cover at 20 *l.*
fhould pay one hundred ‡.

would

---

* The farms are exceedingly fmall, and each farmer almoft keeps a
brood mare.

† Who would venture to breed at fuch an expence?—A tax upon ftallion
horfes would undoubtedly be a very great ftep towards improving the
breed of horfes in general.

‡ The improvement upon horfes in the prefent mode of ferving mares
along the fea coaft, 20 miles north of Liverpool, has taken place for thefe
30 years paft, fo as now to be one-third more in fize and bone, and better
fhaped; and if the prefent breed had been then exifting, would have then
fold for double the price in any market, not faying any thing of the advanced price they have fold for of late.   A tax upon travelling ftallions, if
ever fo fmall, would much difencourage the breed of horfes, and farmers
would be keeping ftallions for their own ufe, of any breed that may fall
into their hands, and the ftallions that now travel the country have moftly
fome merits in them, either for fize, bone, or good fhape, or of fome particular good breed; and the light breed of middling fize and bone are the
moft ufeful horfes for the ftage coaches, and mail coaches, poft horfes, &c.
and many other purpofes that will not bear a high price: the rifk of mif-
fortune is fo great upon horfes that are employed in that bufinefs, and
they will equally ferve the purpofe, as well as one of a higher price, and
often much better; and I have been informed by a gentleman above feventy
years of age, who lives 20 miles eaft of Liverpool, that he has obferved
that the breed of horfes has much improved every feven years for half a
century paft; likewife by a gentleman that is a dealer in horfes, who now
lives in Liverpool, but who was born in a field country, that he has had
perfect knowledge of that country for thirty years paft, and that the breed
of horfes that are now in being there, are as good again as they were thirty
years fince.   I cannot help lamenting that more attention is not paid to
cows being ferved with bulls of good breed, and fuch as would beft fuit
the diftrict, as trials of different breeds might be made with little or no
expence more than the prefent mode.   If fuch a fpirit could be generally
excited in the diftrict that is 20 miles north of Liverpool, and in almoft
every other diftrict in the county, fave the Filde and about Prefton, Lancafter, and Hornby Holme, for the breed of cattle is much the fame as it
was thirty years fince, for little or no attention is paid to the breed, neither
large nor fmall, fo the cow has a calf.—*Mr. Harper.*

Accidental

" Would the produce of such a tax be less than 50,000 *l. per ann.* throughout Great Britain ? By the above tax, the farmer's stock, in the horse line, would in a few years become of infinitely more value. Fewer, being stronger, would be equal to his work, our cavalry better mounted, and a greater sum would annually be returned by foreign nations to this country, for the superior and fine horses we should then be able to export."—In this northern district, and mountainous country, the land is more particularly exposed, and its produce more uncertain. Therefore experiments cannot be made with equal advantage as in the more southern parts of the kingdom.

Many of the lands in this county, are suitable, and would pay well, to breeding *. An improved stock, as before hinted, would return the greatest profit.

Accidental Experiment in the Year 1792. *By the same Farmer.*

I had a heifer calved in the field, and it was some time before she was fetched home, which was not before the calf had suckled itself, by which means she would never give her milk to be milked by hand, for which I put calves to her. After she had had two for the butcher, I then put two young calves on her for rearing, which were on her about ten weeks, and then weaned ; at which time they were better calves than those of four months old, reared in the customary way, that is by poor milk, with the addition of water, meal, &c. The calves did not run on her constantly they were only turned on her at milking time, morning and evening ; and each of them suckled about one half of her milk, as near as could be judged ; and the calf that went on her first in the morning, went last on in the evening ; and they are now two years old, and both in calf, and better beasts by 20 *s. per* head than those reared in the customary way, and equally of as good a breed ; so, for the time coming, I shall conclude one quart of milk, suckled by the calf itself from the beast, to be as good as two of the same quality given any other way ; for it is more natural, nourishing, and strengthening to the calf, while young, and supports it to be of stronger body, and straighter limbed. If such a spirit for rearing calves could be brought forwards with the help of such bulls as would best suit the district, he breed of cattle would soon be much improved, and with a benefit of upwards of 20 *per cent.* more than the present mode.

My opinion is, that if a medal, or a small premium, was to be given to the breeder or farmer that could shew the best stock of horses and cattle of his own rearing, it would greatly encourage the breed of both more than a tax.

* " I think not ; if the land of the county was managed as it ought to be, it would soon become too valuable for breeding.
" The lands in this county, in the southern parts especially, are rented too high for breeding."—*J. W.*

There

There has certainly been a degree of attention paid to the breed of horfes at leaft, for this half century paft, in this county. An attentive obferver on this head remarked, that within the fpace of thirty years, horfes have doubled their value in real goodnefs of quality; whilft the horned cattle, inftead of a progreffive improvement, have been upon the decline. Mr. Bakewell has made the Lancafhire breed the bafis of his improvements.

Oxen have been made ufe of formerly, but always upon a contracted fcale. Horfes at prefent are univerfally preferred for hufbandry bufinefs. The paved roads of this diftrict do not agree with the feet of oxen.

An attentive farmer will make his horfes pay more profit for their keep, than it is poffible for the ox, though this is urged as a ftrong argument in favour of preferring the ox. For if circumftances.permit the farmer to breed ftock, he works them from two years of age, to five or fix, and then fells them off. If the farmer do not breed, the procefs fhould be the fame, to purchafe young cattle, which the eafy and flow operations of agriculture admit to grow and improve. When matured, they become fit for the carriage, road or field, and will then fell, if properly felected, at an advanced price, and fo as to afford a profit for their maintenance, befides the work gained; beyond what is in the limited power of an ox, to gain in weight of carcafe.

On this important fubject, the following obfervations by Mr. Henry Harper merit to be attended to. They arofe from a confideration of the comparative eftimate between horfes and oxen, in the Suffex Report, p. 82.——Mr. Harper's fentiments are as follow:

" I am no advocate for horfes in preference to oxen; but prefer that mode in which bufinefs can be done with moft eafe and leaft expence.

" I have on my farm fome ftrong heavy land as any in the kingdom, and fome as light.—Three horfes, with the allowance of two bufhels of oats per week each horfe, are able to plough an acre a day in the heavieft and ftrongeft land (if ever broke up before) and plough it to any depth from four to eight inches

at

at a proper feafon of the year.—When a feeond ploughing is neceffary, two horfes will be fufficient to plough one acre and a half per day in the fpring or fummer months, and by which there is a fpare horfe, for harrowing in the feed, or any other extra work.—I plough fingle, or the horfes abreaft, as fuits the nature of my work the beft.

" The average work done upon the heavy and light foils on my farm, with a three-horfe team is feven ftatute acres per week the year through, which, at 7 fhillings per acre, is 49 fhillings per week *, and have a fpare horfe eight weeks in the year out of this team.

" My ploughs are the common fwing ploughs with caft-iron mold-boards, of different degrees of ftrength, according to the nature of their work and land under tillage. Single or double wheels may be ufed with thefe ploughs, as occafion requires, and drawn by a chain fixed to the axis of the plough.

" The following is the calculation of the firft purchafe, and keep of three horfes for one year :

	£.	s.	d.
Three horfes, at £. 25 each - - -	75	0	0
Harnefs for ditto, at £. 4. 4s. each - - -	12	12	0
Oats, at 6 bufhels per week, for 6 months -	19	10	0
Oats, at 3 bufhels per week, for 6 months -	9	15	0
Hay for fix months, at 1 l. 1 s. - -	27	6	0
Grafs and green crops for fix months, at 15 s. per week - - - -	19	10	0
Wear and tear of two ploughs, per annum - -	3	3	0
Wear and tear of horfe-gear, per annum - -	1	5	0
Horfe-fhoeing, at 10 s. 6 d. each horfe - -	1	11	6
Farrier - - - - -	0	15	0
	170	7	6
Prime-coft, &c. of ox-team, as ftated in the Suffex report - - - -	147	0	0
In favour of the ox-team, balance - -	£. 23	7	6

* *N. B.* Mr. Harper obferves above, that in fecond p' ughings they are able to plough 1½ acre per day ; therefore he averages feven acres the year through, and allows nothing for lofs of time by bad weather, ima- gining the two acres per week fufficient for that purpofe.

My

	£.	s.	d.
* My horfe-team will earn 49 s. per week per annum	127	8	0
Profit on two young horfes each per annum, befides eight weeks reft for one horfe, or any extra work	2	0	0
	129	8	0

The ox-team will earn 30 s. per week   £. s. d.
for nine months † .   -   -   -   54   0   0
Profit on the oxen   -   -   -   8   0   0

	£.	s.	d.
	62	0	0
	67	8	0
Balance in favour of the ox, firft purchafe - -	23	7	6
Neat balance in favour of the horfe per ann. - £.	44	0	6

" The above ftatement is what a horfe-team will do on my farm, and I think may be done upon any farm in England, where they have proper implements and properly applied."

Thus Doctors difagree in Opinions !

But fince Mr. Harper's management of a horfe-team is fo good, might not an ox-team under his management be alfo more productive ?

Sect. 4.—Hogs.

Pork is not an article of great confumption with any clafs of people in this county. The application of the beft and moft farinaceous kinds of potatoe being chiefly for the food of man, the refufe alone, and the coarfer kinds; fuch as ox-noble, champion, and Surinam ‡, are given to the cows, horfes, and poultry, and to the hogs which may be kept on the farm, which feldom amount to above four.

---

* An allowance is made of two acres every week to make up deficiencies for the whole year, as before ftated.

† The earnings of the ox-team, as well as the earnings per acre of the horfe-team, is according to the calculation made in the Suffex Report.

‡ It is fuppofed moft of thefe coarfe kinds have been raifed from the feeds of the Surinam, and of which they are only varieties, indeed they bear ftrong refemblance to the Surinam, in leaf.

The

LANCASHIRE HOG

with a mixture of the Chinese and the Wild Boar

The idea of hogs being numerous in a potatoe country is very natural; but the fact is not fo: few are bred here, and thofe few that are kept are bought from itinerant drovers from Shropfhire, Yorkfhire, Chefhire, &c. Pork does not feem to be a favourite food with any clafs of people in this county, though more is ufed than formerly. In fhort, the potatoes generally grown by the lower people are of the beft farinaceous kinds; which they are particularly nice in, and confume in their families, or fell to advantage in the market. Some gentlemen and farmers, who grow the ox-noble and other coarfer potatoes, ufe them in general for cows, horfes, and poultry, fcarcely any one keeping more than three or four hogs, which, however, are kept in good condition, and in fome degree fatted with the help of potatoes, but are fatted off at laft with damaged fhip's wheat, India corn, &c. which can often be procured upon reafonable terms from the *corn warehoufes.* Boat loads of ox-noble potatoes are brought to Liverpool from Chefhire, which are bought up for the ufe of cattle, &c.

The ftock of fwine are in general purchafed from herdsmen who travel about the country, and who bring them from Chefhire, Shropfhire, Wales, ahd Ireland. Mr. Ecclefton, however, has a breed between the wild boar and the Chinefe, which have very light and fmall bellies. Upon the fame food, Mr. Ecclefton thinks, they will yield one fourth more flefh than either the large Irifh or Shropfhire. Their fize is but fmall, weight from 10 to 15 fcore, generally about 12 fcore. Mr. Wakefield has the fame breed: an engraving of one of which accompanies this report.

Pigs fhould, during the ftage of their growth, be regularly turned out to graze, where there is a conveniency. This, befides the advantage of grafs, which is nutritious and helps digeftion, by the frefh air and exercife, caufes a difpofition to take their reft, and fleep after a meal, contributes to their cleanlinefs, and renders their flefh of fuperior flavour.

SECT. 5.—*Rabbits.*

THERE are fome lands along the coaft, employed as rabbit-warrens; but thefe animals make excurfions into the adjoining lands, and commit depredations upon the corn: they are all
capable

capable of cultivation; moſt of them poſſeſs marle, either below
their ſurface, or within reach, and are not at all inferior to
Bootle Marſh.

It is a faĉt, however, that neither cows nor ſheep will produce
ſo great a profit as rabbits will afford, on that land which is ſuit-
able for them. Their ſkins, when in ſeaſon, are nearly as valuable
as their carcaſe, and they are prolific to a proverb. A gentleman converted a traĉt of land into a warren, which anſwers well.

Sect. 6 —*Poultry.*

THE Filde is the principal diſtriĉt in this county which
keeps a ſurplus ſtock of poultry. Poulterers alſo colleĉt
the chief part of what is brought to the Ormſkirk market on
Thurſday, from the cottagers and farmers, and retail them out
again at the Liverpool market on Saturday.

On Martin Mere, are turned a number of flocks of geeſe,
on a certain day, brought from different parts of the county.
Theſe flocks are ſo marked, as again to be known. Upon
this Mere they continue till about Michaelmas, and on this
water they can find ſufficient of food for their ſuſtenance from
the different graſſes, aquatics, fiſhes, and infeĉts. The proprietor of the water claims half of the ſtock that remains alive
for their ſummer's keep.

Sect. 7.—*Pigeons.*

A GREAT difference of opinion is entertained in regard
to the utility or the diſadvantage of keeping pigeons. In general, however, it is acknowledged that their dung, in ſo far as
it can be procured, is of the greateſt importance to the farmer.

Sect. 8.—*Bees.*

THESE laborious and uſeful infeĉts, have not been hitherto treated with that degree of attention they merit. The produce of their labour is not only pleaſant, but nutritious; and
before the introduĉtion of ſugar, by the diſcovery of America,
honey muſt have been in high eſteem, by enriching the flavour

of

of many articles, which have only yielded to the introduction and superabundance of sugar *. The wax too is an useful article, and valuable in many of the arts, in which it makes a considerable part of the composition. It is almost incredible indeed, how much can be afforded in the consumption to which it is frequently applied, that of wax lights.

Bees seem to require as little attention to their well-being, as can well be conceived. A straw built cell, with very small accommodation, is what is commonly sufficient, and for which those industrious creatures, in a short space of time, generally repay 10 *per cent.* upon the capital advanced. The pastures from which they gather their rich stores, seem not the least injured; or, in other words they collect and deposit in their cells, and which comes out afterwards either wax or honey (whether by any process of their own, will not, on this occasion, be investigated); a substance, which, if not collected by these industrious creatures would be a loss never to be regained.

These considerations have induced many to contrive methods to preserve their lives, at the expence of their stores, by collateral and other devices in the application of different boxes. These schemes, seemingly humane, have proved in the issue certainly cruel, as a lingering, instead of a speedy death, must be termed so. Too often a bare subsistence for the winter is collected, and if part of that is plundered, the remainder, after a short subsistence, leaves the legal possessor to famine. There-

---

* It is in the memory of a person (*a*), now living, that a family on the borders of the south east part of the county made a complaint, that their bees had not afforded sufficient honey for common use, and that they had been under the necessity of purchasing half a pound of sugar to supply the deficiency in one year.

The surveyor, when a boy, recollects that at the return of the wake (an annual festival, always highly celebrated, by procuring a few superfluities to cheer their friends, who might call upon them), a consultation was held, in a certain family, whether a pound of sugar was to be added to the articles intended to be purchased, which was decided in the negative, and another pound of beef was added to the bill of fare, instead of the pound of sugar.

N. B.—Tea not then introduced.

(*a*) Mr. Titus Hibbert, Manchester.

fore,

fore, if plunder be legal, immediate deſtruction, by fire or
ſulphur, is the greateſt humanity

An accident happening to a hive of bees, belonging to
Thomas Dugdale, of Walton, 1794, the honey was taken,
and after being cleared from the combs was weighed, which
amounted to the aſtoniſhing quantity of 18 lb. in the ſpace of
twenty-one days after ſwarming.

Mr. Lowas, a clergyman in this county, is at preſent employed in
deviſing ſome means to ſave the lives of theſe hitherto devoted and indul-
trious inſects ; together with ſome uſeful experiments and improvements,
which, when ſufficiently aſcertained, will be preſented to the public.

CHAPTER

CHAPTER XIV.

RURAL ECONOMY.

Sect. I.—*Labour.*

THE price paid for different kinds of labour, varies more
in this county, than probably in any other in the kingdom.   An ingenious correspondent observes, " that the rate of
wages is in proportion to the distance of townships from the
seats of manufacturers; *e. g.* at Chorley the wages of a common labourer 3 *s.* with ale; at Euxton 2 *s.* or 2*s.* 6*d.*; at
Ecclefton 1 *s.* 6*d.* or 2 *s.*; at Mawdfley and Bifpham, I am
told you may get them in harvest time, for 1 *s.* 2*d.* and 1 *s.* 4*d.*
in Wrightington the price of labour was lower two years ago,
than the last mentioned sum, and does not now exceed it."

Under this head it may not be improper to give the following statement of different prices of labour, &c. at two periods;
taken by the surveyor after a residence of thirty years in
a village where no manufactory has yet been introduced—
namely, *Walton,* near *Liverpool,*

A comparative Price of Labour, and other Articles, in the courfe of thirty years, taken April 1791

	In the year 1761.			In the year 1791.		
	£	s.	d.	£	s.	d.
Head-man fervant wages per ann. - - -	6	10	0	— 9	9	0
Maid fervant - -	3	0	0	— 4	10	0
Mafons and carpenters, per day - - -	0	1	2	— 0	2	2
Labourers wages † - -	0	0	10 1s. 6d. 1792,	0	1	8‡
Mowing per acre - -	0	3	0	— 0	5	0§
Thrafhing wheat per fcore	0	5	0	— 0	7	6
Barley and beans - -	0	2	6	— 0	4	0
Oats - - -	0	1	8	— 0	2	6
Taylors wages per day and food - -	0	0	6	— 0	1	2
Thatcher per day - -	0	1	0	— 0	2	0
Butcher for killing and cutting up a pig - -	0	0	8	— 0	1	6
Ditto calf, and felling carcafe	0	1	0	— 0	2	6

* At the fame time was taken the number of inhabitants, under their various denominations and occupations ; number of horfes, cows, &c. in each village ; quantity of grain grown, &c. a copy of which was lodged in the parifh chelt (the furveyor being churchwarden that year) in hopes that more ingenious fucceffors in that office might improve upon the hint, and occafionally regifter peculiar circumftances or events. This was done without knowing that the Prefident of the Board of Agriculture was then engaged in a fimilar work over the whole kingdom of Scotland ; which he underftands will be completed in the courfe of the year 1794.

† The hours in fummer fhould be from fix to fix, allowing half an hour at breakfaft, and one hour at dinner; but the labourer in general now comes, or rather leaves home to go to his work, about feven o'clock in the morning, nor continues his labour till the hour of fix, as was the practice 30 years ago—but calculates the time to be taken in his walk home, that he may arrive at the hour of fix. In the winter the hours of labour muft of courfe be curtailed, as are yet, in fome places, the wages—but this practice, of late, is become lefs general.

‡ And an attempt to raife them in the fpring of 1793 to 2 s. per day ; but the calamities, which came on at that period, produced a great change, and every effort was made to procure employment for the induftrious.

§ Eight yards to the rod.

In

<table>
<tr><td></td><td colspan="3">In 1761.</td><td colspan="4">In 1791.</td></tr>
<tr><td></td><td>£.</td><td>s.</td><td>d.</td><td></td><td>£.</td><td>s.</td><td>d.</td></tr>
<tr><td>Butcher for killing a cow, and felling carcafe *</td><td>0</td><td>2</td><td>0</td><td>—</td><td>0</td><td>5</td><td>0</td></tr>
<tr><td>Price of good cart horfes</td><td>10</td><td>0</td><td>0</td><td>—</td><td>25</td><td>0</td><td>0</td></tr>
<tr><td>Pair of men's fhoes</td><td>0</td><td>3</td><td>6</td><td colspan="4">the fame perfon 7 s.</td></tr>
<tr><td colspan="8">and advanced the end of that year to 7 s. 6 d.</td></tr>
<tr><td>Sett of horfe-fhoes</td><td>0</td><td>1</td><td>0</td><td>—</td><td>0</td><td>1</td><td>8</td></tr>
</table>

Carpenters work—price of feveral particulars ufed in Agriculture.

<table>
<tr><td></td><td colspan="3">In the year 1761.</td><td colspan="4">In the year 1791.</td></tr>
<tr><td></td><td>£.</td><td>s.</td><td>d.</td><td></td><td>£.</td><td>s.</td><td>d.</td></tr>
<tr><td>Large cart 7 feet 3 inches, wheels 5 feet 2 inches high, with flakes, complete, twice painted (to the carpenter)</td><td>5</td><td>0</td><td>0</td><td>—</td><td>9</td><td>4</td><td>0</td></tr>
<tr><td>Ringing a pair of wheels</td><td>0</td><td>18</td><td>0</td><td>—</td><td>1</td><td>15</td><td>0</td></tr>
<tr><td>New axle-tree, and work</td><td>0</td><td>4</td><td>0</td><td>—</td><td>0</td><td>6</td><td>6</td></tr>
<tr><td>Wheel-barrow, and trundle</td><td>0</td><td>5</td><td>0</td><td>—</td><td>0</td><td>12</td><td>0</td></tr>
<tr><td>Plough</td><td>0</td><td>7</td><td>0</td><td>—</td><td>0</td><td>11</td><td>0</td></tr>
<tr><td>Harrow, 3 feet 6 inches</td><td>0</td><td>3</td><td>6</td><td>—</td><td>0</td><td>5</td><td>6</td></tr>
<tr><td>Pair of homes</td><td>0</td><td>0</td><td>6</td><td>—</td><td>0</td><td>0</td><td>9</td></tr>
<tr><td>Spade fhaft</td><td>0</td><td>0</td><td>4</td><td>—</td><td>0</td><td>0</td><td>6</td></tr>
<tr><td>Common five barred gate</td><td>0</td><td>5</td><td>0</td><td>—</td><td>0</td><td>10</td><td>0</td></tr>
<tr><td>Ladders, 15 ftaves, per ftave</td><td>0</td><td>0</td><td>4</td><td>—</td><td>0</td><td>0</td><td>4</td></tr>
<tr><td>Ditto, from 15 to 30 ftaves</td><td>0</td><td>0</td><td>0</td><td>—</td><td>0</td><td>0</td><td>6</td></tr>
<tr><td>Swipels, ftens, and fets for carts</td><td>0</td><td>0</td><td>2</td><td>—</td><td>0</td><td>0</td><td>6</td></tr>
<tr><td>Wheat per bufhel</td><td>—</td><td>—</td><td>—</td><td>—</td><td>0</td><td>7</td><td>6</td></tr>
<tr><td>Barley</td><td>—</td><td>—</td><td>—</td><td>—</td><td>0</td><td>3</td><td>6</td></tr>
<tr><td>Oats</td><td>—</td><td>—</td><td>—</td><td>—</td><td>0</td><td>2</td><td>6</td></tr>
<tr><td>Beans</td><td>—</td><td>—</td><td>—</td><td>—</td><td>0</td><td>4</td><td>6</td></tr>
</table>

* The journeymen butchers in Liverpool, about thirty-three years ago, flaughtered at the following prices : a bull 2 s.; a cow 1 s.; a fow 6 d; a fheep 1½ d.; a calf 3 d; of the laft, about twelve were one day's work; alfo one fcore, or two dozen of fheep, were a day's work. The prices are now doubled,

Wheat-

In 1761.					In 1791.		
	£.	s.	d.		£.	s.	d.
Wheat-ftraw per load	0	5	0 per ftone of 20℔.		0	0	3½
Barley-ftraw per thrave -	0	0	2½	—	0	0	6
Oat-ftraw per thrave -	0	0	5	—	0	0	9
Butter per lb. from 5 d. to 8 d.			- - -	from 8 d. to 1			.
* Sweet milk per quart -	0	0	1	—	0	0	1
Eggs, two and-three for 1 d.			from 1 d. to 2 d. per egg.				
In the winter of 1794 -			- -	3 d. per egg.			

N. B.—Expended upon the poor from Eafter

					£	s.	d.
1760 to Eafter 1761	-	-	-	-	22	3.	2¼
From Eafter 1790 to Eafter 1791	-	-			115	14	1

There have been twenty additional houfes built in the fpace of time betwixt 1761 and 1791.

The above ftatement feems to confirm the opinion of fome, " that the poor-rates increafe as the price of labour advances;" which in fome places, (as appears from the anfwers given to the agricultural queries) have been as high as nine, eleven, and thirteen fhillings in the pound.

Piece Work, or by the Great.

Making new fence, ditch, hedge, bank, feven fods in height, backing, and covering with thefe fods, planting quickfetts, bearding, from 1 s. 6 d. to 2 s. per rod.

Cutting hedges, opening and fcouring the ditches, putting frefh earth to the quicks, from 8 d. to 14 d. per rod.

Delving or trenching with dung, one fpit or fpade deep, 10 d. to 1 s. 3 d. two fpits 1 s. 6 d. to 1 s. 8 d. per rod; digging for peas and beans 6 d. and 8 d. per rod; double gutters 1½ foot deep, 4½ d. to 6 d. per rod (of 8 yards); common fpade gutters 1½ d. to 2 d. per rod; feighing two yards deep, or if under, 2½ d. to 3 d. the folid yard.

Mowing from 3 s. to 4 s. per ftatute acre; reaping from 3 s. 6 d. to 5 s. per acre.

---

* To what caufe is the unvaried price of this valuable article to be attributed ? It is flattering to the modern improvement of meadow lands, by the growth of various graffes, formerly hardly known, and by the cultivation of this land in general, if this induftry and attention may have effected fo effential a benefit.

2   Thrafhing

Thrafhing is done fometimes by the thrave, and fometimes by the bufhel—the price generally paid by the piece is about one twentieth of the value of the grain, or one bufhel of the grain thrafhed at every fcore.

Effects of Piece Work.

In many cafes Piece-work is defirable, as it encourages a fpirit of difpatch, and, in confequence, proves a fource of benefit to an induftrious labourer; at the fame time it is a temptation to labourers to over-work themfelves, which ought to be avoided. Gentlemen who employ a number of workmen together, fhould be extremely guarded, not only in their choice of men, but alfo a proper infpector; fince wherever one is difpofed to loiter, either by telling his ftory to divert his companions, or by any means caufe an intermiffion of labour, all the company muft of courfe become lifteners, and the fpace of five minutes, in the company of twelve, is equal to the lofs of a whole hour's labour of one individual. Nor is this the whole of the evil. Bad examples are contagious. Thofe who might be formerly induftrious, become by flow fteps more indolent. The contagion fpreads wider, and the evil increafes.

Sect. 2.—*Provifions.*

Butchers meat, like other articles in this county, varies in price. It is generally deareft towards the fouth and fouth eaft, many cattle being driven from the northern part to fupply thofe diftricts; but ftill, it is there generally more than a penny per pound under the London market-price. Corn, at Liverpool, is always above the London price, nearly one fhilling per bufhel, as appears by the returns publifhed. In thofe parts of the county where oat-meal is chiefly ufed for bread, &c. when enquiry was made after the price of provifions, the firft anfwer was univerfally the price of oat-meal, the ftaff of their life.

At Manchefter market, October 9, wheat fold that day from 33 to 34 s. per load, as it is termed, or fack, of 16 fcore. Oats 33 to 34 s. per load of 9 Winchefter bufhels. Beans 30 s. per load of 5 Winchefter bufhels. Potatoes 4 s. 6 d. to 5 s. per load of 12 fcore, 12 lb. wafhed; unwafhed, thirteen fcore.

Fine

Fine flour 36 *s.*; feconds 34 *s.*; thirds 26 and 28 *s.* per 12 fcore; oat-meal 36 and 37 *s.* per load, of 12 fcore.

No barley at this market.

Cheefe from thirty to fifty fhillings per cwt.

The price of provifions, unlefs the feafons are very unfavourable, is more likely to fall than to advance, if trade continues to ftagnate.

In eftimating the prices of meat, due regard fhould be paid to the qualities of the meat, different values of the different joints of meat of the fame quality, and the different feafons of the year—veal being generally cheapeft when beef and mutton are the deareft.

In the year 1793, the prices of beef might be from 3 *d.* to 5 *d.* per lb.; mutton from 3 *d.* to 6 *d.*; and veal from 3 *d.* to 6 *d.* per lb.

The writer of this paid the whole of that year $4\frac{1}{2}$ *d.* per lb. for his meat, all (except lamb) weighed together. The average confumption in his family 100 lbs. weight per week.— The meat was of the very beft quality, and of which the top part of the buttock, provincially called a *round*, a fhoulder of veal, and hind quarter of mutton, almoft univerfally made three ftanding joints every week in the year—in the other joints fometimes the butcher and fometimes the purchafer was accommodated.

Whence the Markets are fupplied.

The principal fatting diftricts in this county are from Claughton to Hornby, a rich pafture there called the Holmes, and from thence through that fertile vale as far as Kirkby Lonfdale * ; alfo fome gentlemen's parks, and private inclofures, but the whole of thefe amount to a mere trifle, compared to the confumption requifite. The deficiency is made up from the counties of Weftmoreland, Durham, Yorkfhire, Lincolnfhire, Derbyfhire, and Shropfhire; the principality of

* A calculation has been made by two perfons, who feem competent for fuch work, by knowing every farm, its fize, and nearly the number of ftock kept on each; and their account is 2,000 head of horned cattle, and 5,000 of fheep.

Wales,

Wales, the kingdoms of Ireland and Scotland, are alfo applied to, to fupply the county of Lancafter with beef and mutton. The county itfelf furnifhes a very fmall proportion of the bread and meat actually confumed there. Nay, the poultry and the pigeons are fupplied from diftant parts. Befides what comes from the Filde (the only diftrict in the county which, with a few trifling exceptions, has any furplus of ftock) the Liverpool market has fupplies from Chefhire, Wales, Ifle of Man, Scotland, and Ireland. Manchefter alfo receives great fupplies from Chefhire, Derbyfhire, Lincolnfhire, and even Nottinghamfhire. Eggs of courfe muft be purchafed, and come from the fame quarters, and fome at a greater diftance, packed up in cafks. Some come even from Kendal, and Penrith *.

S E C T. 3.—*Fuel.*

COALS in general abound, and are cheap, infomuch that a fmall family may fupply itfelf with fuel for about 30 fhillings per annum. No wood confumed, but the refufe of fhip-carpenters, and other workers in wood. Peat from the different moffes, is an article of fuel in the vicinity of thofe places, but feldom without the addition of fome coals. Faggots, which were formerly an article of confumption among the bakers of fea-buifcuit, and other bread in Liverpool, has for fome years been difcontinued; coal is preferred, and by experience find it more advantageous. This circumftance is well worthy the attention of other towns, as the faggots require large room, and may be attended with danger.

* Some of the eggs fold at Manchefter are packed up with layers of ftraw between every row of eggs, about ten thoufand in one cart. The man brings two carts, and comes every fortnight during the feafon that a fufficient number can be collected; which is chiefly done by women who travel the country with mugs and other articles, which they exchange for eggs in Cumberland, &c. There are two or more higglers (*qu.* egglers?) who follow this practice, befides the old man who gave the information above, and who was counting them out to the huckfters. Few eggs are broken by the carriage. The man is four days upon the road. It feems the collectors of the eggs are paid 6 *d.* per hundred for collection.

B b   CHAPTER

CHAPTER XV.

POLITICAL ECONOMY.

SECT. I.—*Roads.*

MR. YATES obferves, that there is a greater length of
roads in this county, in proportion to its extent, than in
any other county in the kingdom, and of fo little public utility,
that many might be fpared; and he alfo remarks, that if early
exertions had been made upon this head, land fufficient in
value, might by that means have been obtained, to have kept
the whole remaining roads in proper repair.

This opinion may have been too fanguine, and the beft opportunity for accomplifhing fo defirable a work, may have now
paffed. But, no doubt, much advantage to the county might
yet be obtained by proper exertions, if roads, that at prefent are
of little public utility, were ftopped, the lands fold, and the cafh
arifing appropriated to fupport the remainder.

In proof, however, of this affertion, of the vaft length of
roads in this county: the parifh of Goofnargh contains
3703 acres, and the length of the roads in that parifh is nearly
forty miles, befides three miles of bridle road, and three miles
of road repaired by certain individuals.

ɔnThe townfhip of Walton, near Liverpool, which only contains 1988 ftatute acres, has a public road two miles and a half
in length; parochial roads, eleven miles two furlongs, befides
occupation roads.

In the northern and north-eaftern parts of the county, materials for making roads are found upon the fpot, the lime-
ftone, which, when broken, binds together, and makes an
excellent road; but in the midland and fouthern parts, the
materials, except what the rivers afford, are brought from the
Welfh and Scotch coafts, and at confiderable expence.

Thefe are Boulder ftones, and they are not broken, but

† paved.

paved. The whole expence of which is from 1 *s.* 2 *d.* to 2 *s. per* fquare yard, according to the diftance of the materials to be carried. Two quarries of pebbles have lately been difcovered. Copper fcoria or flag, from two works, Ravenhead and Liverpool, have been fuccefsfully tried. This article makes an excellent fide road to the pavements, and is preferred to pavement both by the horfeman and drivers of carriages.

Great exertions have been made of late years, at very confiderable expence *, to improve the roads; the effects of which are very apparent, both upon thofe which are public and parochial.

Pavements are the moft expenfive, and moft difagreeable of all roads, but we have no other material that will ftand heavy cartage.

Near Warrington, Mr. Kerfoot, who undertook the management of the Prefcot and Manchefter turnpikes, has made admirable roads with the copper flag.

Mr. Holt, who is furveyor for one parifh, made an attempt with copper flag, but it is difficult to get the flag fufficiently broken.

The town of Liverpool is a great enemy to turnpikes. There are only three toll-gates within eight miles of it, none within four.

Commercial and manufacturing towns have *a fyftem* of throwing every poffible burden upon the land.

The toll-bars here, as well as in other parts, from private views and intereft, are improperly placed—fhould they not in each act, be placed in the moft advantageous fituations for the benefit of the road by ftrangers, commiffioners appointed for that purpofe, and private intereft totally be laid afide ? Moft of the great towns have had fufficient intereft to place the toll-bars at fome miles diftance from them. The toll-bar on the road to the fouth from Liverpool is placed at 5 miles diftance from the town. Would it not be a fair claufe in the general

* So great, that at the time when Mr. Yates took his furvey, about ten years fince, the average paid through the county, was not lefs than eighteen-pence in the pound.

act

act of parliament, when the inhabitants of a town object to a bar being placed near to the town, that they fhould engage to keep in repair the road from the town to the bar (which is in general the moft expenfive part of the whole) without receiving the leaft benefit from the money collected? The diftance the bars are placed from the great towns in this county, is almoft the fole caufe of the wretched condition of the turnpike roads.

An ingenious road-maker in the neighbourhood of Warrington, has of late exploded the common *convex* form, and adopted that of *one inclined-plane*; the inclination juft fufficient to throw off occafional water. By this alteration he finds that a road becomes more durable; for when it is convex, all heavy carriages ufe the center of it, and keep in the fame track; therefore the center is foon deftroyed, and the fides feldom ufed: but when a road has only one fmall inclination, the whole furface is ufed, for, in this cafe, you will feldom fee two carriages take the fame line.

With refpect to improvements, an ingenious gentleman obferved, that the tolls in general ought either to be raifed, or the number of bars increafed, in order that the public at large might contribute a proper quota, for their eafe in travelling, by the improved ftate of the road, and the farmer, &c. of courfe eafed; and candour muft allow, that the facility, expedition, and fecurity of travelling over the roads, in their prefent ftate, is worth more than double the money paid for this convenience. Some method fhould be devifed to eafe the labourer, and lay the burthen upon the traveller. The tenant has frequently been charged with an unexpected tax, amounting to 4 or 5 s. in the pound, upon a fhort leafe, when a fine has been levied; and though, in the iffue, this clafs receives as great benefit as any other, ftill fome method fhould be devifed to eafe thofe contingent poffeffors, by more heavily taxing the travelling ftranger.

Under this head, the indulgence fhewn to the mail coaches in their exemption from tolls, merits reprehenfion.

In

In the firft place, the objeÆ is too trifling and mean, for the interference of government. It is alfo an encroachment upon private property, and upon a capital, the intereft of which was expeÆed to be paid upon the credit of certain tolls, with an accumulating furplus, to repair the damage done to the roads by the paffing of thefe carriages—and with the remaining portion, to liquidate the principal advanced to accommodate the public in the execution of thefe undertakings *. But here is a check upon thefe fpirited endeavours by encroachment. If the price at prefent paid for the carriage of the mail be not fuffi-cient, it fhould be increafed by an addition taken from the com-mon ftock.

But the profit arifing to the proprietors of mail coaches is at prefent great. The furveyor was informed lately of the fol-lowing ftatements as proofs of the affertion: The receipts of the mail coach from London to Liverpool, and backward, amounted, in the courfe of one month, in the fpring of the prefent year, to twelve hundred pounds †. The other ftate-ment is—that the profits arifing from the length of one ftage (10 or 12 miles) were lately fold, and transferred, for the neat fum of three hundred pounds.

As this bufinefs is, at prefent, conduÆed in a fpirited manner, and probably the moft expeditious, fafe, and neat conveyance in the world, the proprietors and conduÆors of fuch public ac-commodations, ought to have, not only certain, but handfome profits. What is here objeÆed to, is the infringement upon private property. And if thefe tolls were not allowed, they would be charged at laft upon the paffenger, upon whom they ought certainly to fall.

But again, the tolls allowed to be taken for this fpecies of carriages, if they were even extended to the mail coaches, are

* Mail coaches prevent much-travelling poft—confequently injure the toll-bars moie ways than one.

† Thefe ftatements are here given as related to the furveyor, and are not to be depended upon as authenticated faÆs. When a fubjeÆ becomes a topic of converfation, there are generally fome grounds for the affertions, which fhould however be received, till fully authenticated, with diffidence.

not fufficient for the damage done by them, in proportion to the rates paid, and the damage done by other carriages to the roads.

The weight of a mail coach, loaded with paffengers and parcels, may be near two tons, the heavy coach nearly three tons.

The effects of four horfes, fcampering and pulling with all their might, are very injurious to the roads; for, after the ftones have been nearly difplaced by this exertion of the horfe-feet (very different to the effect of a road-horfe), followed by a heavy carriage, fupported and dragged upon four narrow wheels, every obftruction is difplaced by the violence of the motion. The flow pace of a waggon, moving upon a nine-inch furface, or a heavy-loaded cart, under two or three tons burden, upon fix-inch wheels, makes a comparifon ftrongly in favour of thefe carriages.

Again, the tolls arifing from many turnpikes are very in-fufficient to maintain the roads. The townfhip of Walton, at the prefent juncture, is meeting the truftees of the public road, which runs through that diftrict, with not a lefs fum than four hundred and thirty pounds, befides ftatute labour, upon a length of two miles and a half; whilft the fame townfhip is burdened with other roads of the length of eleven miles two furlongs and a half, as before obferved.

All the townfhips through which this turnpike paffes are, at prefent, contributing their aid, and that to a degree in fome places not a little burdenfome to both tenant and free-holders; of which the townfhip of Aintree is a ftrong ex-ample *.

The propereft roads for this part of the county, particular the neighbourhood of Manchefter and Liverpool, and all the coal diftrict, would be roads fimilar to thofe of France and Flanders: a pavement in the center, made of large fragments of granite (which might be imported from Scotland, at no

" The remarks which the furveyor makes on turnpike roads, are worthy the obfervation of the honourable Board, for they are ftubborn facts."

great

great expence) on each fide of this pavement fhould be a gravel
road, of the beft material the country could afford, and made
of fufficient breadth, and kept in fuch good repair as to induce
all light carriages to prefer it to the pavement in the center: I
prevailed upon the furveyor of this townfhip to make an experiment of backing up a high pavement with copper flag
(fcoriæ) fome years ago, and to cover it with the loofe fandy
rock of the country. It is now the beft part of the turnpike.

* In addition to the above, it may be neceffary to ftate that
from the vaft increafe of carriage in this county, and the general ufe of waggons, carts, &c. with *exceffive weights*, it is
become almoft impoffible, by any means, and at any expence,
to fupport the public roads. The climate is wet, the foil
foft, the ftone and gravel found in the county are not hard or
lafting, and the only materials that have ftrength and durabi-
lity are the paving ftones imported from the coafts of Wales,
at the heavy price of fix fhillings per ton. Some of the
turnpike roads in the neighbourhood of Manchefter, paved
with thefe ftones, coft from £. 1500 to £. 2000 per mile.
Fortunately thefe ftones were exempted, in the act of laft feffions, from the tax on ftone exported, or Lancafhire muft have
been at once reduced to a miferable fituation. Yet the oblig-
ing the floops employed in collecting and carrying thefe
paving ftones, to take out *coaft difpatches and certificates* (as
in the cafe of coal and falt exported coaftwife in Scotland) by
the delays and expences hereby incurred, adds a very confiderable impoft on thefe articles, without any benefit to the revenue; and this hardfhip is too apparent not to be immediately remedied. The legiflature has at all times been
wifely provident, not only to authorize and require the making
of good roads, (which are unqueftionably the firft improvements in any country) but alfo to enact rules for their *prefervation*.

The encouragement of broad wheels, or rolling wheels, or
carriages fo conftructed, " *as to enable them to carry great*

* By T. B. Bayley, Efq.

weights,"

weights," was always a doubtful meafure. Experience now puts it out of queftion, that thefe *heavy weights* foon deftroy the beft conftructed roads, and exhauft all the common materials for their reparation. The turnpike trufts are thus more deeply involved in debts inextricable, or difproportionate tolls are levied to fupport an injurious fyftem, oppreffive to the country, and ruinous even to the carriers and waggoners, who purfue this *miftaken* fcheme of bufinefs. '

The reftraints of weighing machines are found to be expenfive, partial, and quite ineffectual ; and the only remedy for this great and increafing evil, is that pointed out near thirty years fince *by the Rev. Henry Homer*, of Warwickfhire, in his " Enquiry into the Means of *preferving* the public " Roads," printed at Oxford 1767, viz. " *fuch a conftruction* " *of carriages as will* oblige *them to carry LIGHT loads.**"

In fupport of this fcheme of *preferving* our roads, and of *faving* an immenfe fum of money now annually fquandered away, there is a vaft body of evidence in the excellent volumes of the Statiftical Account of Scotland, and the Surveys of Counties, reported to the Hon. Board of Agriculture. Thefe all prove what is ftated in the Survey of *Cumberland*, page 48, that " two horfes, yoked in *fingle* horfe carts, will draw as much " as three horfes yoked in one cart."

The general ufe of *fingle horfe* carts would be a vaft faving in the number of horfes kept for labour, and of hay and corn expended in their maintenance, would be gainful to the carriers, &c. and would preferve the roads, and take off the increafing and oppreffive burden of taxes now raifed (but ineffectually) for their fupport. The exemptions from toll, or being weighed, given to carriages employed in *hufbandry*, are in moft places (efpecially in the neighbourhood of great towns) very injurious to the roads, and not warranted by any fair analogy of taxation, which ought equally to affect all who are benefited by it, and by what mode foever. The regulation refpecting the flat conftruction of wheels, fo as to prefent an

---

* See alfo Mr. Jacob's Treatife on Broad Wheels, &c. (Diliy, 1774) and Annals of Agriculture, vol. xviii. p. 178.

even

even furface to the road, alfo of the flat tire and counterfunk nails, are ill defined in our general turnpike act, and worfe in practice. By the 16th Geo. III. c. 39. fect. 2. it is enacted, " that fix inch wheels fhall be deemed *flat*, as fhall not de- " viate more than one inch from a flat furface." This figure will fhew on how few points of a good road this *flat* wheel will bear, and how more injurious it muft be than a common narrow wheel.

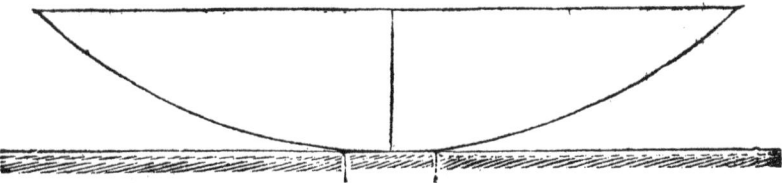

The truth is, this *important* fubject is little underftood, or attended to, and requires a careful revifion. This may fpeedily be hoped for from the exertions of the Board of Agriculture. The act of the laft feffions (34th Geo. III. c. 74.) feems to be formed on the old miftaken principle of *fixing* what from fituations and times muft ever be various and fluctuating, viz. *the price of labour.* Parliament cannot fix its maximum or its minimum. The higheft price for compofition for a team per day is fixed now at 6 s. whereas eight or ten fhillings per day is paid in many parts of this county for the labour of fuch team. The better way would be to leave the prices to be annually fettled, and publifhed by the magiftrates of the feveral divifions when they appoint furveyors of the highways.

The ftatute duty, or compofition, taken from labourers renting under *five* pounds a year, had better be *wholly abro- gated.* It is an odious burden, is rarely collected, and with difficulty and expence (in counties like this) not to be conceived.

The relief propofed to thefe poor people in the fifth fection

would

would be attended with fo much lofs of time and money to them, and fo much inconvenience to the furveyors of the highways, that it is plain it *never can* have any operation.

We have generally, during the two laft years of diftrefs, omitted to call for the ftatute duty or compofition from this defcription of inhabitants; and by law to free them from the obligation would, at this feafon efpecially, be a juft and a *popular* meafure, which I earneftly recommend.

With refpect to *turnpike roads*, in this, as in other counties, there is not a due regard paid to the general public convenience, in making the roads in the moft direct lines, or on the eafieft levels. And if there fhould be a neceffity of making fhort turnings or elbows, they fhould be at leaft twice as broad as in the other parts, that the *thill* horfe may have not the whole load to draw, whilft the others are turning; and fuch place fhould be made level as poffible:—but hills of even one furlong in length, are fometimes fo fteep as to require an additional horfe for that fhort fpace; and if the road is often to be paffed, the additional expence of keeping one horfe, one might imagine, need only be pointed out to obtain their removal. There is indeed fcarce any part of the kingdom that might not have been laid out, fo as to fuperfede the neceffity of ufing that badge of barbarity a chain to a waggon-wheel. — " When " our defcendants fhall become more fenfible than we feem to " be of the advantage of level roads; no expence will perhaps " be confidered too great, to remove an evil, which nothing but " habit could render fufferable *."

The obligation on *parifhes* to repair roads by prefcription (fee Hawkins's Pleas of the Crown, part ift, page 202) wants to be limited by *ftatute*.

This plan has, under the late great change of circumftances, brought an intolerable burden on the *parifh* of Manchefter, which includes a great extent of country, and an immenfe

* Herefordfhire Report.

population

population, to repair ways, hitherto little known or used, but now *become public streets in the town itself.*

I would propose a clause to limit to 30 years back the proof of use and repair by the parish, and to allow parishioners to be competent witnesses on either side.

Amongst the various objects of enquiry, and to which answers have been returned to the Board of Agriculture, there is none of more general importance than the state of the *public roads.* As a measure of *national police*, this has *not* hitherto been sufficiently attended to by the legislature : the 13th Geo. III. chapter 84, commonly called the *General Turnpike Act*, is very inadequate and greatly mistaken in many of its provisions. The introduction of *turnpikes* into England is of a very late date; they were at first established for the confined limits of local convenience; and have gradually been so multiplied and extended, as to form almost an universal plan of communication through the kingdom, supported by a *public tax* of vast amount.

In this *national* view of the subject, connecting the public convenience and prosperity, and the large sums raised throughout the kingdom to render the general communication easy and certain, it cannot be denied, that the revision of the general law, the adoption of a better system for making roads, is now become necessary ; a system founded, *not* on speculations of mere local or private convenience, and as affecting particular towns, districts, or even counties, but on the more extended considerations of general intercourse and common benefit. In fact, we may observe in every part of England the *jobbing trade*, as it respects turnpike roads, very industriously pursued. The old course is generally followed, however circuitous or difficult.

Heavy carriages are still to be dragged over the summits of steep hills, formerly scarcely accessible to the pack-horses of the country, whilst the easy and obvious levels of the adjoining vallies are overlooked. Happy would it be for the country, if all plans, for *turnpike roads* were settled in the manner described by Dr. Anderson in his " View of the Agriculture of *Aberdeenshire*," p. 135.

As

As turnpike bills have been ufually too much confidered as *private* bills (though none are of more *public* concern) the committees of the Houfe of Commons have ufually done little more than confirm the agreements of the meetings previoufly held in the country, in which perfonal and local interefts frequently fuperfede a due confideration of general benefit. The experience which thefe committees have had on various occafions of this felfifh fpirit, has produced fome very falutary " orders relating to bills for making turnpike roads."

To enable thefe committees more accurately to judge of the propriety of future application for making new or *amending old* turnpike acts, I would fuggeft another ftanding rule and order ; viz.

" That, together with the eftimate of expence, and the account of the money fubfcribed (as ordered by the 3d rule) there be delivered to the committee an exact plan of the propofed road, on a fcale of   to a mile, fhewing its connection with the neighbouring towns ; together with an accurate *fection* of the whole line of road."

S E C T. 2.—*Canals.*

I N granting new bills for cutting navigable canals, care fhould be taken by the legiflature, that lime or manure be carried upon low terms. The introduction of wealth, in confequence of fuperior cultivation, by the means of manures, &c. will introduce the carriage of more bulky articles, and foon repay the proprietors the trifling indulgence. A gentleman obferved, that, as a certain portion of land was loft to the community, either for tillage or pafture, by cutting canals, care ought to be taken in the banks to preferve as much grafs as poffible, by burying the rubbifh under ground, and applying the beft foil to cover the furface of the banks ; trifling as fuch an object may be, as canals are daily increafing, the amount, in the iffue, would be fomething, and would repay to the public a fum fufficient for the general attention requifite.

The many canals already begun, and intended, have had
                                                    confiderable

considerable effects both upon the agriculture, manufactures, and general state of the country *.

The Sankey canal was the first inland navigation in the kingdom, and was opened in the year 1756; after which the Duke of Bridgewater's canal; and then the Leeds canal, as far as Wigan, were completed. The canal from Kendal, through Lancaster, to Westhoughton, is a great undertaking, ten miles of which are already completed. The Bolton canal, already begun, the Rochdale canal intended, with the navigable rivers Mersey, Douglas, Ribble, Wyre, and Loyne, render the carriage of heavy articles, through the internal parts of the county, more easy and less expensive, than where such channels of conveyance are not found. They have no small effects upon the agriculture of the county, in conveying dung, lime, and other articles, into parts whither, without their assistance, they could hardly have been transmitted; as also upon the manufactures, by the conveyance of coal and raw materials, the gross weight of which would have been too expensive upon carriage by land.

S E C T. 3.—*Fairs.*

IN the year 1780, August 2, a fortnight fair was established at Harrington, near Liverpool, opposite St. James's church, by the north-country graziers, to shew fat cattle and sheep, which was encouraged by the butchers in Liverpool and the neighbourhood. Accommodations for the cattle and sheep were effected by Mr. Samuel Sandys, who then held upwards of forty Cheshire acres of land, which was appropriated to the purpose, and was continued every fortnight until the 12th of February, 1783; when it was removed to Kirkdale, for convenience to the butchers in Liverpool; during which period there were exposed for sale 39,160 sheep, and 8,309 head cattle, and upwards: in the year 1781, at one show, in September, were 1,489 sheep and 279 head cattle; and in October, 1782, at another fair were 1,691 sheep and 343 head cattle, which was thought very con-

---

* Particulars of what business is done in each, and their connections with the trade of Liverpool, will be given in the intended history of that town.

siderable

fiderable.   After the diffolution of this market, Mr. Sandys had applications from cow-keepers for the land, which was much improved by the pafturage of drovers, fheep, and cattle; alfo by the quantity of manure which was collected from his ftall-feeding thirty head in fhades built on the premifes, which was declined on removal of the fair; therefore Mr. Sandys propofed finding milk cows, and keeping them at grafs or hay for 5 *d.* per head per week at his own rifk, or keep *their* cows at 4 *d.* per week at their rifk; and when any cow declined fo much as not to pay the farmer, he had a frefh cow found, or an abatement in proportion to her decreafe: this mode kept the land in high condition from the quantity of dung collected on the eftate, &c.

The old eftablifhed fairs are not here noticed, fince they are publifhed in the ufual kalendars of thefe things.

S E C T. 4.—*Weekly Markets.*

THERE are faid to be twenty-fix market-towns in the county, which are fuppofed fufficient for the inhabitants, becaufe in every little village or hamlet of houfes, there are retailers of the different articles, which are of daily confumption, in great abundance.   The two large towns, Manchefter and Liverpool, have each two market days every week; but of late years, butchers meat, garden-ftuff, and a number of the neceffary articles of life, are expofed to fale, and may be purchafed any day in the week, Sundays excepted.

S E C T. 5.   *Commerce.*

THE foreign commerce carried on by the county of Lancafter, is extremely confiderable, but its nature and extent does not come within the object of this Report. It is material, however, to collect information refpecting that great branch of the trade of the county, which interferes with its agricultural interefts, namely, the importation of corn; fome idea of the extent of which, may be formed from the following ftatements of the quantity of corn imported to and exported from Liverpool alone, in the years 1791, 1792, &c.

WHEAT,

WHEAT, FLOUR, &c. imported into Liverpool during the years 1790, 1791, and 1792, from Foreign Parts.

Year	WHEAT.		BARLEY.		OATS.		BEANS.		RYE.		PEASE.		Wheat Flour.*			Oatmeal.		
	Qrs.	Bu.	Qrs.	Bu.	Qrs.	Bu.	Qrs.	Bu.	Qrs.	Bu.	Qrs.	Bu.	Cwt.	Qu.	lb.	Qrs.	Bu.	lb.
1790	68,260	4	14,404	5	204,154	1	17,492	4	1,288	0	69	6	22,000	2	11	6,874	7	33 Bolls.
1792	164,311	1	8,213	1	171,591	7	4,467	1	55,520	2	17	1	51,654	0	25½	41,320 Cwt.	5 Qr.	0 lb.
1793	8,369	0	19,489	4	223,737	3	27,821	1	2,576	3	1,387	3	6,489	2	9	9,125	1	8

WHEAT, and other GRAIN, imported into Liverpool, coastwise, in the years 1791 and 1792.

Year	Wheat.	Barley.	Meal.	Rye.	Oats.
1791	31,273	63,305	46,927	2,290	9,667
1792	71,236	6,597	35,375	3,456	38,797

GRAIN exported coastwise.

Year.	Wheat.	Barley.	Meal.	Rye.	Oats
1791	30,912	6,597	2,942	3,975	12,292
1792	5,148	3,052	4,197	3,440	16,078

* Notwithstanding the quantity of fine flour, both imported, and at present consumed in this county, Robert Winstanley, a miller, now resident in Liverpool, aged about seventy, says, that he remembers the first dressing-mill fitted up in this county, which was at Walton, near Preston; and which, at the time of a scarcity, was threatened to be demolished by the mob, for dressing fine flour to feed the rich (a); and on which occasion the mill was converted to another use, to which it is applied to this day. That afterwards he, with an elder brother, who had learned the art of dressing fine flour, fixed up a dressing machine at Bootle-mills, near Liverpool; which was the second mill in the county, where fine flour was ground upon blue stones (b), and afterwards dressed through a cloth. Before this, the flour was dressed, and sifted at home in sieves, after being ground at the mills, and the fine (or London flour as it was then termed) was purchased, on extraordinary occasions, at the grocers shops, made up into pounds, similar to the present mode of making up sugars in blue papers. These facts are confirmed by a letter from Sir H. Hoghton, Bart. to the surveyor, dated Dec. 1, 1793; and that mill was then the property of his uncle, Sir H. Hoghton.

- (a) There was more waste then, than there is now; too great a proportion of flour being left in the bran: the improvements in this art have since caused it to be more effectually extracted, and that to a degree, as to grind almost the whole of the bran with the flour.

(b) The flour first made use of for grinding fine flour in preference to the grey-quarry stone, was the blue boulders, sawn and cemented together; but this stone acquiring a polish after some using, was insufficient; afterwards the French stone, a porous, loose, hard stone, was introduced, and has been since used.

This

This extract from the Cuftom-Houfe books, with both the imports and exports, will fhew the great confumption of grain in this county, and how inadequate the land, in its prefent ftate, is to the fupply of its inhabitants.

The exportation of corn is trifling; and, except upon the weftern borders of Yorkfhire, upon the eaftern boundaries of Chefhire, and fome parts of Derbyfhire, the corn imported into Liverpool is chiefly for the confumption of Lancafhire.

The average of the Liverpool import of grain
for the laft three years is   -   -   78,980
The average of the Norfolk export for laft three
years is   -   -   -   -   -   63,046
Liverpool import at £. 1. 4s. - is 173,211
Norfolk export at - £. 1. 4s. - is 138,701
————
34,510 more · value
imported into Liverpool, than exported from Norfolk.

There are about 1,500 tons of fea-bifcuit manufactured for the different veffels that fail from the port of Liverpool, which is eftimated to take about 60,000 bufhels of wheat, and to require the labour of about fifty men with boys. This is about the average in the year 1792.

Obfervations by Major Atherton.

Mr. Kent, in his Report to the Board of Agriculture, having ftated that the four Norfolk ports export as much corn as all the reft of the kingdom put together, and having entered into an accurate detail from the Cuftom-Houfe books, it occurred to me that a comparifon between the exported produce of the county of Norfolk and the corn imports of the town of Liverpool might eventually be of fome ufe to the Board, I have therefore taken fome pains to obtain intelligence upon this fubject; and here lay the refult of my inquiries before them. The Liverpool prices were taken from the average prices of one of the firft houfes in the corn-trade belonging to the port. More difficulties have however arifen than I was at firft aware of, and I am confident that it is ftill extremely defective;
fuch

fuch as it is, however, it may be the caufe of further enquiries from thofe who are better calculated than myfelf to examine a matter which is certainly of high importance. The difference of weights and meafures produce endlefs and almoft infuperable difficulties in enquiries of this nature.

TABLE

TABLE of Comparison between the IMPORTS of the Town of *Liverpool* and the Exports of the County of *Norfolk*, on the Average of the Years

1790, 91, and 92.

Exportation from NORFOLK, according to Mr. KENT's Report.

			£	s.	d.
Wheat -	63,046 Quarters,	at 44s. per Quarter -	138,701	4	0
Wheat Flour	37,135 Do -	at 56s. per Do -	103,978	0	0
Barley -	360,380 Do -	at 24s. per Do -	432,456	0	0
Malt -	90,271 Do -	at 42s. per Do -	180,542	0	0
Rye -	14,056 Do -	at 25s. per Do -	17,570	0	0
Pease -	13,361 Do -	at 28s. per Do -	18,705	8	0
Beans -	15,148 Do -	at 24s. per Do -	18,177	12	0
Vetches -	73 Do -	at 30s. per Do -	109	10	0
Rape Seed -	2,423 Do -	at 36s. per Do -	4,361	8	0
			914,601	2	0

Deduct 15,389 Qrs. Oats imported more than *exported*, at 17s. per Quarter - 13,079 13 0

Total neat Exports - 901,521 9 0

Deduct not brought into Liverpool Account,

	£	s.	d.			
Malt -	180,542	0	0			
Vetches -	109	10	0	}		
Rape Seed	4,361	8	0			

£ 185,012 18 0 .... 185,012 18 0

	£	s.	d.
	716,508	11	0
Liverpool Import -	643,312	8	9
Balance in favour of Norfolk Exportation - £	73,196	2	3

1790, 1791, and 1792.

Importation of LIVERPOOL, according to Mr. HOLT's Report, calculated at Norfolk Prices, from Foreign Parts only.

			£	s.	d.
Wheat -	80,313 Qrs. -	at 44s. per Quarter -	176,688	12	0
Wheat Flour -	26,714 Cwt. 1 qr. 24 lb. which equal 5,311 qrs. 43 lb. reckoning 45 lb. to a Winchester bushel	at 56s. per Quarter -	23,271	3	0
Barley -	14,035 Qrs. -	at 24s. per Quarter -	16,842	0	0
Oats -	201,494 Qts. -	at 17s. per Quarter -	171,269	2	0
Rye -	3,128 Do -	at 25s. per Quarter -	3,910	0	0
Beans -	16,593 Do -	at 24s. per Quarter -	19,911	0	0
Pease -	458 Do -	at 28s. per Quarter -	641	0	0
Oatmeal -	7,111 $\frac{107}{115}$ Loads, at 23s. 3d. per Load -		8,267	10	9

Total Foreign Importation - £ 420,800 7 9

Balance of Grain imported Coastwise, after deducting the whole Exports of the Port,

			£	s.	d.
Wheat -	33,224½ Qrs. -	at 44s. per Qr. -	73,093	18	0
Barley -	30,125 Qrs. -	at 24s. per Do -	36,150	0	0
Oats -	10,047 Qrs. -	at 17s. per Do -	8,539	19	0
Wheat Flour	37,581 Qrs. 4 Bush. -	at 56s. per Do -	104,728	4	0
			222,512	1	0

Foreign Importation, as above - 420,800 7 9

£ 643,312 8 9

SECOND TABLE of Comparison between the IMPORTS of the Town of *Liverpool* and the EXPORTS of the County of *Norfolk*.

Exportation of the County of NORFOLK, according to Mr. KENT's Report,

On the Average of the Years 1790, 91, and 92.

		£.	s.	d.
Wheat -	63,046 Quarters, at 44s. per Quarter -	138,701	4	0
Wheat Flour -	37,135 Quarters, at 56s. per D° -	103,978	0	0
Barley -	300,380 Quarters, at 24s. per D° -	432,456	0	0
Malt -	90,271 Quarters, at 42s. per D° -	180,542	0	0
Rye -	14,056 Quarters, at 25s. per D° -	17,570	0	0
Peas -	13,361 Quarters, at 28s. per D° -	18,705	8	0
Beans -	15,148 Quarters, at 24s. per D° -	18,177	12	0
Vetches -	73 Quarters, at 30s. per D° -	109	10	0
Rape Seed -	2,423 Quarters, at 36s. per D° -	4,361	8	0
		914,601	2	0
Deduct 15,389 Quarters of Oats imported more than exported, at 17s. per Quarter -		13,079	13	0
	£.	901,521	9	0

	£.	s.	d.	
Deduct Malt -	180,542	0	0	Not brought in
Vetches -	109	10	0	Liverpool ac-
Rape Seed -	4,361	8	0	count -
£	185,012	18	0	

	£.	s.	d.
	185,012	18	0
£.	716,508	11	0

Importation of LIVERPOOL, according to Mr. HOLT's Report, valued at Liverpool Prices,

On Average of the Years 1790, 91, and 92, from Foreign Parts only.

	£.	s.	d.
Wheat - - 80,313 Quarters equal 523,181 Bushels of 70 each, at 7s. 2d. reckoning 57 lb. to a Winchester Bushel -	187,473	9	8.
Wheat Flour - 26,714 Cwt. 1 qr. 24 lb. at 38s. 6d. per pack of 280lb. or 16s. 6d. per Cwt.	20,568	15	5¼
Barley - - 14,035 Quarters, at 32s. per Quarter -	22,456	0	0
Oats - - 201,494 Quarters, at 18s. per Quarter -	181,344	12	0
Beans - - 16,593 Quarters, at 33s. per Quarter -	27,447	11	9
Rye - - 3,128 Quarters, at 32s. 4d. per Qr. -	5,056	18	8
Peas - - 458 Quarters, at 42s. per Quarter -	961	16	0
Oatmeal - 7,111 $\frac{107}{246}$ - at 25s. 4d. per Load of 5 Bushels -	9,007	17	0
	454,317	0	0¼

Balance imported Coastwise more than *exported*.

	£.	s.	d.
Wheat - - 33,224 Quarters equal 216,433 $\frac{62}{70}$lb. 70lb. at 7s. 2d.	77,555	9	7
Barley - - 30,125 Quarters, at 32s. per Quarter -	48,200	0	0
Oats - - 10,047 Quarters, at 18s. per Quarter -	9,042	6	0
Wheat Flour - 37,581 Quarters, 4 Bushels, at £.3. 1s. 1d. 7d. per Quarter of 448 lb. -	115,719	14	0¼
Total neat Import - £.	704,834	10	2¼

Mr.

I have many powerful reafons to believe that the account of the corn, &c. imported into Liverpool, as ftated in the firft report, is erroneous, that the importations are much greater, and, at any rate, that it is extremely defective. The article malt is entirely omitted; now in the year 1794, from the 1ft of January to the 28th of April, fay four months, there were imported at Liverpool coaftwife, 9,070 qrs. 7 bufhels, or 72,567 bufhels Winchefter. In the fame four months were imported 105,726 bufhels of barley, and 46,072 bufhels of big, coaftwife; and 44,635 bufhels of barley, from 5th January to 5th April 1794.

In the week ending March 12th 1795, Liverpool imported from Ireland only 45,627 quarters of oats, befides 1,699 quarters from Englifh and Scotch ports. In all,

47,326 Quarters - - - at 20 s. -	£. 47,326 0 0
at prefent price - - 24 s. . -	£. 56,791 4 0

Imports of the town of Liverpool at Liverpool prices -	£. 704,834 10 2
Do - - - - at Norfolk prices -	643,312 8 9

Superior value at market or profit to corn-dealers, after deducting freight, infurance, intereft, &c. &c. - - £. 61,522 1 5

Norfolk exports annually,

Wheat - - 63,046 Quarters, at 44 s. - -	£. 138,701 4 0
Wheat Flour - 37,135 Quarters, at 56 s. - -	193,978 0 0
Beans - - - 15,148 Quarters, at 24 s. - -	18,187 12 0
In all - 115,329 Quarters - - - -	£ 250,856 12 0

Liverpool imports annually,

Wheat, Foreign 80,313 Quarters, - at 57 s. 4 d.	£. 187,473 9 8
Wheat Flour, Do - - 26,714 1 14 at 38 s. 6 d.	
per peck - - - - - -	20,568 15 5¾
Wheat - - Coaftwife 33,224 Quarters, at 57 s. 4 d. -	77,555 9 7
Wheat Flour, Do - - 37,581 4, at £. 3. 11. 7 d.	
per Quarter - - - - -	115,719 14 0¹
Beans - - - 16,593 Quarters, at - -	27,447 11 9
	£. 428,765 0 6

It

It appears from hence that there is a market in beans and wheat alone to the annual value of £.428,765. 0*s*. 6*d*. more than the entire diftrict produces, and for nearly £.177,908. 8 *s*. 6*d*. more than the whole county of Norfolk exports.

Is this, or is it not, an argument for converting the unprofitable grafs land of this county (of which, I am forry to fay, the quantity is immenfe) into good cultivation ?—Is it a reafon for marling ?—Will it pay for manure and tillage ?

Beans, managed in the Kentifh manner (fee Ann. Agr. vol. ii. p. 70, &c.) are amongft the beft of preparations for wheat. Few or none are grown in Lancafhire and Chefhire, and thofe few univerfally broadcaft.

At this moment, wheat is felling at Liverpool and at Warrington for 10*s*. and 10*s*. 6*d*. per bufhel, of 70 lb. The common preparation for wheat in this diftrict is a fummer fallow, even upon light fands.

For clover the fale is ready, and the confumption profitable; and it ought to precede wheat upon all barley lands.

No county can produce better barley, or in larger quantities, when properly cultivated. It always fetches a fair price, being not only ufed for malting, but made into bread, either by itfelf, or mixed with wheat.—The great miftake of this diftrict is, fowing it too late, and fowing it after wheat.

There is no better or furer land for turnips in England than in this county; and there is every where a good market for them, where it is not convenient to eat them off the land with fheep.

Marle and manure are every where to be had in great abundance.

The material obftacles to improvement are tythes, poor rates, and the immoderate wages to be obtained at the manufactories.

A quarter of wheat-flour in Norfolk, weighing 448 lb. is worth   -   -   -   -   -   -   £.2 16 0

At

At Liverpool the fame weight and quantity may average - - - - - - - - £. 3 1 7
                                 2 16 0

            Superior value at Liverpool - c 5 7
            or per bufhel, nearly - - 0 0 8½

Say 7 _d._ per bufhel wheat, equals 8 ½ _d._ flour; and fay 32
bufhels, of 70 lb. each, is an average crop upon a ftatute
acre; the fuperiority of market is, per acre, then £. 0 18 8
          Double it, per Chefhire acre - c .8 8

          Advantage per Chefhire acre - £. 1 17 4

          At five quarters per acre, it is - £. 2 6 8

There are vaft tracts of land in this county, rented at lefs
than 40 fhillings per Chefhire acre, capable of producing the
above quantity. This country then has three powerful in-
centives to improvement,

        Marle, Manure, and Markets.

I have heard it confidently afferted that this diftrict (the
counties of Lancafter and Chefter) do not fupply the confumption for more than fix weeks in the year, and that the
county of Lancafter in particular, does not grow more grain
than would feed or be confumed in it in two weeks.

I am fenfible of the great imperfection of many of the
above ftatements; and poffibly there may be many notorious
errors in the calculations: I hope, however, the fubject will
be taken up by fome perfon whofe talents are equal to the
tafk.

Such are Major Atherton's intelligent remarks on the table
inferted in the original report; but that the beft authority
might be gained, application has been made by the Board to the
Infpector General for an account of three years, which is
alfo inferted, and in addition to it the value, at the Liverpool
prices.

An ACCOUNT of the EXPORT and IMPORT of all Sorts of CORN and FLOUR, Foreign and Coastways, at the Port of Liverpool, in the following Years.

	1791.				1792.				1793.				1794.			
	FOREIGN PARTS.		COASTWAYS.		FOREIGN PARTS.		COASTWAYS.		FOREIGN PARTS.		COASTWAYS.		FOREIGN PARTS.		COASTWAYS.	
	Imported.	Exported.	Brought in.	Carried out.	Imported.	Exported.	Brought in.	Carried out.	Imported.	Exported.	Brought in.	Carried out.	Imported.	Exported.	Brought in.	Carried out.
	Qr. Bu.															
Wheat	150,311 0	—	31,273 0	9,192 0	8,137 4	3 0	71,336 0	5,148 0	96,149 7	—	16,461 0	13,013 0	122,112 2	Cwt. qr. lb.	51,499 0	11,020 0
D° Flour	Cwt. qr. lb 45,534 3 5	Cwt. qr. lb 2,147 0 0	—	—	Cwt. qr. lb 6,489 3 9	Cwt. qr. lb 4,283 0 14	—	—	Cwt. qr. lb 33,468 3 5	4,693 3 11	—	3,434 0	511 3 25	1,819 2 0	—	—
Barley	9,947 6	15 0	18,300 0	9,597 0	19,250 0	27 3	23,308 0	3,051 0	10,385 4	28 0	37,315 0	—	15,459 2	—	34,124 0	6,775 0
Beans	41,137 0	87 0	—	—	28,607 7	94 0	—	—	6,045 7	—	—	—	17,223 5	461 0	—	—
Oats	170,930 0	311 4	9,670 0	123,273 0	239,008 3 Bolls.	392 0	2,719 0	21,650 0	125,230 5 Bolls. lb.	155 0	3,623 0	9,610 0	228,647 4 Bolls. lb.	150 0	9,491 0	2,439 0
Oatmeal	771 7	302 0	—	—	7,300 36	52 3	—	—	6,201 13	36 2	—	—	2,611 19	750 0	—	—
Pease	72 3	34 0	—	3,975 0	1,557 6	41 1	16 0	3,416 0	484 6	1 5	—	7,236 0	3,041 5	7 0	17 3	3,361 9
Rye	6,975 2	Cwt. qr. lb. 338 3 21	—	—	2,576 3	—	—	—	9,046 2	—	—	—	2,273 2	—	—	—
D° Flour	—	—	—	—	5,695 0	—	—	—	—	—	—	—	—	—	—	—
Indian Corn	329 0	—	—	—	—	—	—	—	—	—	—	—	—	—	—	—
Foreign Wheat	—	—	—	—	—	Cwt 8,523 0 26	—	—	—	Cwt. 2,315 0 1,286 3 9	—	—	—	22,584 7	—	—
D° W. Flour	—	Cwt. qr. lb. 4,724 0 1	—	—	—	2,668 3	—	—	—	—	—	—	—	—	—	—
D° Beans	—	204 4	—	—	—	459 1	—	—	—	44 4	—	—	—	—	—	—
D° Oats	—	—	—	—	—	1,559 4	—	—	—	—	—	—	—	—	—	—
D° Pease	—	—	—	—	—	436 4	—	—	—	43 6	—	—	—	25 1	—	—
D° Rye	—	—	—	—	—	7,076 6	—	—	—	955 1	—	—	—	—	—	—
British Malt	—	—	45,937 0	26,942 0	—	—	450,975 0	3,197 0	—	—	26,693 0	130 0	—	—	21,558 0	130 0

Inspector General's Office,
Custom House, London,
June 13th, 1795.

THOMAS IRVING,
Inspector General.

The material originally positioned here is too large for reproduction in this reissue. A PDF can be downloaded from the web address given on page iv of this book, by clicking on 'Resources Available'.

TABLE of COMPARISON between the
IMPORTS of the Town of *Liverpool*, and the EXPORTS of the County of *Norfolk*.

EXPORTATION of the County of NORFOLK, according to Mr. KENT's Report, on the Average of the Years 1790, 91, and 92.		£. s. d.	IMPORTATION into LIVERPOOL, according to the Account furnished from the Custom House, for the Average of the Years 1791, 92, and 93; and valued at Liverpool Prices.		£. s. d.	£. s. d.
Wheat - 63,046 Qrs. - - at 44 s. per Qr.		138,701 4 —	Wheat - {Imported - 84,866 Qrs. - at 57 s. 4 d. {Brot Coastways, 39,657 ditto - ditto	243,282 10 8 / 113,683 8 —		356,965 18 8 / 23,392 10 6
Wheat Flour - 37,135 ditto. - at 56 s. per ditto.		103,978 — —	Wheat Flour, Imported - 28,597 Cwt. - at 16 s. 6 d.	— —		
Barley - 360,380 ditto. - at 24 s. per ditto.		432,456 — —	Barley - {Imported - 17,194 Qrs. - at 42 s. {Brot Coastways, 20,308 ditto - ditto.	21,110 8 — / 42,092 16 —		
Malt - 90,271 ditto. - at 42 s. per ditto.		189,569 2 —	British Malt, Brot Coastways, 38,848 Qrs. - at 42 s.	— —		63,203 4 — / 81,580 16 —
Rye - 14,056 ditto. - at 25 s. per ditto.		17,570 — —	Rye - {Imported - 6,499 Qrs. - at 32 s. 4 d. {Brot Coastways, 5 ditto - ditto.	10,506 14 4 / 8 1 8		
Pease - 13,361 ditto. - at 28 s. per ditto.		18,705 8 —	Rye Flour, Imported - 113 Cwt. - at 32 s. 4 d.	— —		10,514 16 — / 182 13 8
Beans - 15,148 ditto. - at 24 s. per ditto.		18,177 12 —	Pease - Imported - 701 Qrs. - at 42 s.	— —		1,472 2 —
Vetches - 73 ditto. - at 30 s. per ditto.		100 10 —	Beans - Imported - 12,930 Qrs. - at 33 s.	— —		21,334 10 —
Rape Seed - 2,423 ditto. - at 36 s. per ditto.		4,361 8 —	Oats - {Imported - 179,056 Qrs. - at 18 s. {Brot Coastways, 7,004 ditto - ditto	161,150 8 — / 6,303 12 —		167,454 — —
		923,628 4 —	Oatmeal - Imported - 2,474 Qrs. - at 40 s. 6 d.	— —		5,009 17 —
			Indian Corn, Imported - 1,808 Qrs.			
					£.	731,310 7 10
			Wheat - {Exported - 2,427 Qrs. - at 57 s. 4 d. {Card Coastways, 26,135 ditto - ditto.	7,129 8 — / 26,135 8 —		
			Wheat Flour, Exported - 8,586 Cwt. - at 16 s. 6 d.	— —		33,264 16 — / 7,083 9 —
			Barley - {Exported - 14 Qrs. - at 32 s. {Card Coastways, 5,357 ditto.	22 8 — / 8,571 4 —		
			British Malt, Card Coastways, 1423 Qrs. - at 42 s.	— —		8,593 12 — / 2,988 6 —
			Rye - {Exported - 1,010 Qrs. - at 32 s. 4 d. {Card Coastways, 4,876 ditto - ditto.	1,632 16 8 / 7,882 17 4		
			Pease - Exported - 186 Qrs. - at 42 s.	— —		9,515 14 — / 390 12 —
			Beans - Exported - 305 Qrs. - at 33 s.	— —		503 5 —
			Oats - {Exported - 676 Qrs. - at 18 s. {Card Coastways, 13,977 ditto - ditto.	608 8 — / 12,579 6 —		13,187 14 — / 263 5 —
			Oatmeal - Exported - 130 Qrs. - at 40 s. 6 d.	— —		
					£.	75,790 13 —

Three acres of wheat ſtraw have been ſold for the enormous ſum of ſix guineas the acre of large meaſure by Mr. Harper.

The improved mode of cultivating potatoes has reduced their price of late years, notwithſtanding the conſumption by cattle has been ſo great.   The laws admitting importation of grain prevent the farmer gaining an advance of price when there is a failure of crop ; and the value of corn is, by this means, kept within ſome bounds.   But the methods ſometimes taken, as is ſaid, on the opening and ſhutting the ports, ſtand in great need of regulation.   The only advantage the farmer reaps, is, from additional quantity, never from advanced price ; which is not the caſe in regard to hops, or ſugar, or other articles produced by the ſoil, either at home, or in our colonies.

S ᴇ ᴄ ᴛ. 6.—Of Manufactures.

Mᴀɴᴜꜰᴀᴄᴛᴜʀᴇs have been carried on to a very conſiderable extent in Lancaſhire.

The cotton *, ſilk, and wool †, through all their branches,

---

* The firſt piece of cotton, manufactured from Britiſh growth, was at Mancheſter, from cotton grown in the grounds of J. Blackburne, Eſq. M. P. of Orford, in Lancaſhire ; ſeven yards and a half, of one yard and a half yard-wide muſlin, from four ounces of raw material, raiſed I ſuppoſe in a hot-houſe.   It was a moſt beautiful piece of cloth, propoſed to have been made up into a dreſs, for Mrs. Blackburne, in which ſhe intended to have appeared at Court, June 4, 1793 ; but was prevented by a change of dreſs, occaſioned by the loſs of a relation.

To what a degree of perfection the muſlin manufactory is arrived, the following may ſerve to convey ſome idea.   In the year 1791, a ſingle pound of cotton was ſpun to a fineneſs of ninety-ſeven poſt miles in length : the muſlin, after being ſpun, was ſent to Glaſgow, to be wrought, and after which was preſented to her Majeſty.   The pound of cotton, which, in its raw ſtate coſt 7 ſ. 6 d. coſt the ſum of 22 l. in this ſtage, when it was wrought into yarn only.   It was ſpun by one Lomax, at Mancheſter, upon the machinery called mules.

† Woollens have of late been manufactured without either ſpinning or weaving, and after the manner of hats.

†   from

from the raw material; and thefe leading articles include a number of fubordinate branches or trades, *e. g.* fpinners, bleachers, weavers, dyers, printers, and tool-makers for the different artifts, which; if feparately enumerated, would in the aggregate extend to an amazing amount.

There are alfo manufactories of hats*, ftockings, pins, needles, nails, fmall wares, tobacco and tobacco-pipes, fnuff, earthen-ware, Englifh porcelain; clocks and watches, and tools for the artifts in thefe two branches, not only for the neighbourhood but for all the world; long bows, fteel bows, paper, &c.

There are large works for the fmelting of iron and copper †, of cafting plate-glafs, and the fabrication of blown glafs; the procefs of making white lead, lamp-black, vitriolic acid, and foffil alkali, the refining of fugar, &c.

The feveral modes of accelerating labour have been always ftoutly refifted by the labouring clafs, when the different machinery was firft introduced; but the iffue has hitherto proved a fource, from which not only employment, but the price of labour has increafed, notwithftanding that labour has been fo much abridged.

---

* A patent has been obtained, and a work eftablifhed, to manufacture hats, by machinery; moved by water.

† The confumption of coal at Ravenhead is, feven hundred tons per week; and however deftructive the fmoke may be to vegetable life, it feems more, favourable to animal; fince, in the fpace of fourteen years, notwithftanding between two and three hundred people are conftantly employed in the copper-works there, belonging to the Paris Mine Company, not one perfon, employed in the works, has died. One reafon, why perfons in large manufactories in Lancafhire, do not as frequently die in great numbers as in other counties, is that they have (in general) been *inoculated* in their infancy.

Inoculation is the moft effectual of all expedients for preferving the fhort-lived race of man—many gentlemen pay for the inoculation of the children of the poor in their own neighbourhood.

Saddleworth,

Saddleworth, which borders upon the county, and which formerly only wrought coarfe woollens, has gained lately, and now works, the fine weftern woollen cloths.

A large manufactory for the fabrication of fancy goods, has lately been eftablifhed at Tildefley, by Thomas Johnfon, Efq; where a village has been built fince the year 1780, which had then only two farm-houfes and nine cottages; has, in 1793, 162 houfes, and a new chapel erected. The village contains nine hundred and feventy-fix inhabitants, which employ three hundred and twenty-five looms.

Manchefter being the principal repofitory for thefe manufactures, has become the great center, to whioh not only the country retailers, but merchants, from all quarters of the kingdom, and foreign parts, refort; and this has induced feveral capital woollen houfes to fettle at that town; and this mart is chiefly confined to one ftreet, in which a fingle room frequently lets very high.

The trades and different occupations upon which the maritime ftate depends, have not, on this occafion, been noticed; becaufe they are the fame in all counties where navigation is carried on *.

The good or bad effects which manufactures may have had upon agriculture, is an important queftion, which merits much attention; the anfwers to which, in fome letters, have been concife, and difcharged by one fingle word, *e. g.* one anfwer has been " advantageous;" another anfwer " injurious;" but without either argument or proofs to fupport thefe laconic affertions.

The more extenfive anfwers, however, fhall be faithfully ftated.

Manufactures have wrought a change in the agriculture of the county; the growth of grain is annually and gradually on the decreafe. The importation from foreign countries is, of courfe, upon the advance; the diminifhed ftate of cultiva-

* A fketch of fome of which will be given in the intended Hiftory of Liverpool.

tion is one caufe of this, and the increafing population is another; and by the joint operation of thefe two, the importation of grain and flour, ufed chiefly in this county, is almoft incredible. To prove which, the furveyor has been favoured with extracts from the cuftom-houfe books, faithfully, and with no fmall trouble, collected for this occafion, by Mr. Yates.

The advance of wages, and the preference given to the manufacturing employment, by labourers in general, where they may work by the piece, and under cover, have induced many to forfake the fpade for the fhuttle, and have embarraffed the farmers, by the fcarcity of workmen, and of courfe advanced the price of labour.

The poor rates fall, with equal burden, upon the farmer, as upon the mafter manufacturer; and the manufacturers encourage fettlers, and confequently increafe the number of paupers.

The water is fometimes fo damaged by dye-houfes, and other works, erected upon rivers, as to be rendered not wholefome to the cattle, and deftructive to fifh. The heat neceffary for the bufinefs of printing debilitates the ftrongeft conftitutions.—Damps from obftructed water;—peftilential air from crouded rooms;—effluvia from acids and different preparations;—down from cotton; all operate as peftilences to the human conftitution.

On the other hand, the advantages that have been held forth, have been an increafe of population; as that which conftitutes the riches and ftrength of a country.

Increafe of the value of lands, and alfo of provifions. The farmer particularly has an advance on the price of his cheefe, his butter, his fatted cattle, his milk; alfo ftraw, which, in 1790, fold at the advanced price of 8 *d. per* ftone in the fpring at Liverpool; dearer, probably, than ever was known, even in the London market. Hay is little dearer than thirty years ago, except on extraordinary occafions;—hay is, at prefent, about 8½. *d per* ftone, owing to a flight crop;—thirty years ago 6 *d.. per* ftone.

3   Capitals,

Capitals, labour, ingenuity, and attention are in this county diverted from agriculture §. It is much to be lamented that the Board of Agriculture have not employed some persons of extraordinary talents and superior industry, to examine, in the different manufacturing districts, the actual effects of manufactures upon agriculture *. This county, as Mr. Young himself observes in his most valuable reflections upon this subject, subjoined to his Tour into France, carries on manufactures to a greater extent than any other county in the kingdom, and is at the same time nearly the worst cultivated.

By way of illustrating this remark, which is equally true and important, let us examine the chief articles of cultivation, and the method of management adopted in this great manufacturing and commercial county, where the land is capable of producing every vegetable and every grain in great perfection and abundance :

Beans and Peas.—As preparation for wheat, seldom.—Always broadcast.—Hoed by horse or hand, never.

---

* The following are the observations of a practical farmer upon this important subject.—" From various circumstances it evidently appears, that trade is injurious to agriculture, and in the end to landed property, unless it could be restricted ; for whenever a stagnation in trade happens, the poor rates rise, and the land pays for it. Poor rates and other taxes in West Houghton have amounted this year to 16 s. in the pound. Corn is not so much grown', for though the farmer can get in his grain, he cannot raise hands but at an enormous price to reap it : if mowing corn were more practised, it would be better.

Another farmer says, " Never enquire about the cultivation of land, or its produce, within ten or twelve miles of Manchester ; the people know nothing about it : speak of spinning-jennies, and mules, and carding machines, they will talk for days with you.

" There are people about Ashton that give £. 6 for a summer's grass for horses to work carding engines, and will give from £. 12 to £. 15 for hay and after-grass, that they may not be troubled with cultivating land to hinder them, as they say. If land were attended to, and improved, for ten to fifteen miles round Manchester, as it is in Derbyshire, the lower parts of Yorkshire, Nottinghamshire, &c. it would be as productive as any land in any part of England ; for it all inclines to marle, and is naturally a strong soil, not only fit to carry manure of any kind, but hold it for a sufficient time.

　　　　　　　　Cabbages.

Cabbages.—Plentiful and abundant, and luxurious in gardens; but as an arable crop in fields, unknown.

Turnips.—Never hoed.—Never fed off upon the land with sheep.

Vetches.—Winter vetches unknown.—Summer ones sown when the land will produce nothing else; not eat green, but made into hay.

Fallows.—Seldom ploughed before winter, but kept to starve horses and young cattle.—Green, with couch-grass, in June.

Oats.—Sown perpetually upon the same land, consequently deficient in quantity and quality.

Barley.—Sown in May and June.—Never weeded.

Wheat.—Universally fallowed for, even upon light sands.—Upon clover lay, seldom if ever.—After beans, never.—The bean-stubble is too weedy.—Never weeded or hoed. Though the land is every where admirably adapted to the cultivation of wheat, not a hundredth part grown that ought to be, that the poorer class of people from Lancaster to Preston, Chorley, Blackburne, &c. &c. seldom taste wheat, though they inhabit as good wheat lands as any in the kingdom.

In the vicinity of Manchester, Wigan, Warrington, Ormkirk, Prescot, and Liverpool, there are many large tracts, to which the above assertions will not apply; and every where there are intersperfed both professional men, and gentlemen whose management is correctly just:—I speak of the generality of the county.

There are many just observations upon this subject in Mr. Campbell's Account of the Filde, " the Granary of Lanca-" shire," printed in Ann. Agr. vol. xx. p. 109; they merit general attention, and have more justice than superficial observers would allow.—There needs little to prove the importance of manufactories in a national view; and their effect upon agriculture, theoretically speaking, seem immense, in as much as they form the best and most certain markets:—But, practically speaking, they are baneful to agriculture.

The

The immediate wages to be obtained in the manufactories rob agriculture of its moft valuable fupporters;—the yeoman and the labourer are both tempted from the plough;—all competition is precluded.—Who will work for 1 *s.* 6 *d.* or 2 *s.* a day at a ditch, when he can get 3 *s.* 6 *d.* or 5 *s.* a day in a cotton work, and be drunk four days out of feven?—But their moft deftructive effect are the increafe of the poor rates. In winter many hands are turned out of employment, who muft be fupported by parifh rates; the labourer at cotton muft, when fick or ill or aged, be fupported by taxes levied upon agriculture. — Manufactories encourage fettlers of all defcriptions.—Above 5,000 Irifh were fettled at Manchefter in the year 1787, and I am told that number was afterwards doubled.—The poor laws in this circumftance are extremely defective.—The law decrees, " that if any perfon who fhall " come to inhabit in any town or parifh, fhall be charged with " and pay his fhare towards the public taxes or levies of the " faid town, he fhall be adjudged to have a legal fettlement " in the fame, though no notice in writing fhall be delivered " and publifhed." (See *Burn's Juftice.*)—By way of a commentary upon this law, there is a manufacturer at this time at Prefton, who has refufed to pay his parifh rates and taxes, unlefs they are lowered.

Another evil arifing from manufactories is, the propagation of vice, infubordination, and difeafes.—What elfe can arife from the multitude of people of all defcriptions pent up in printing-houfes, from which it is neceffary to exclude all exterior air, and to keep up an artificial heat, which muft of courfe debilitate the ftrongeft conftitutions?—Add to this, effluvia from acids, paints, minerals, and charcoal.

In the neighbourhood of Bolton, bleaching of the very beft quality in the kingdom is performed; and of late has been introduced by M. Vallete (an ingenious Frenchman) a more expeditious method of bleaching, fo much that a piece of calico which would have required by the cuftomary procefs three weeks in the moft favourable feafon, may now be rendered perfectly white in the fpace of one hour, and that, as it is faid,

without

without the leaſt injury ſuſtained by the cloth. The new pro-
ceſs is ſomewhat more expenſive than the old. And there is
as much ingenuity diſplayed amongſt the artificers in Bolton,
and its neighbourhood, as in any part of the county. Bolton
has been long celebrated.

" Bolton upon Moore market ſtondith moſt by cottons and
cawrſe yorne. Divers villages in the moores abowte Bolton
do make cottons. Nather the ſcite nor ground aboute Bolton
is ſo good as at Byri. They burnè at Bolton ſum canale, but
more ſe cole, of the wich the pittes be not far of. They burne
turfe alſo." *Leland's Itinerary,* vii. *p.* 49.

Upon the ſubject of manufactures, a celebrated agriculturiſt
obſerves " that you muſt not go for agriculture to Yorkſhire
Lancaſhire, Warwickſhire, or Glouceſterſhire, which are full
of fabrics, but to Kent, where there is not a trace of a fabric;
to Berkſhire, Hertfordſhire, and Suffolk, where there is
ſcarcely any. Norwich is an exception, being the only great
manufacture in the kingdom, in a thoroughly well cultivated
diſtrict, which muſt very much be attributed to the fabric
being kept remarkably within the city, ſpreading (ſpinning excepted) not much into the country; a circumſtance that de-
ſerves attention, as it confirms ſtrongly the preceding obſervations. But the two counties of Kent and Lancaſter, are ex-
preſsly to the purpoſe, becauſe they form a double experiment.
Lancaſter is the moſt manufacturing province in England, and
amongſt the worſt cultivated; Kent has not the ſhadow of
a manufacture, and is perhaps the beſt cultivated *."

S E C T. 7.—*Poor.*

Whatever may be the ſtate of the poor, they are
moſt liberally provided for, not only by legal aſſeſſments, but
liberal contributions—when particular ſeaſons, or calamitous
circumſtances, may call forth the humanity of thoſe who, on
ſuch occaſions, give without ſparing. Yet, with all the aid

* Travels through France, by Arthur Young, Eſq. vol. ii. p. 508.

of large affeffments, and liberal contributions, it is truly lamentable to witnefs fuch appearance of poverty, exemplified in nakednefs, dirtinefs, and the different garbs which indicate diftrefs. There are mendicants of all ages and fexes, but more particularly in the country villages; the exerted police of well-governed towns reftrains thefe wanderers.

In brief, it may be afferted, that from appearances, the *ftate of the poor* is not fo comfortable as might be wifhed; and yet from the fums levied and contributed, if properly applied, their fituation might be meliorated.

Friendly focieties feem the guides which point out radical cures for the exifting evils. When a man once gets into the habit of *laying up* in ftore, however fmall the capital, he feels a fatisfaction which ftimulates exertions to increafe his ftock; and that pride of independence which enfues from an enjoyment of the acquifition of his well-deferved, however hardearned fubftance, render his meals fweet, his family regular, clean, and decent, and his fpirits cheered by the fruits of his own labours. Friendly focieties have been the means of caufing all this among many of their members; they are numerous in this county; they are increafing, and ought to be encouraged.

SECT. 8.—*Population.*

LANCASHIRE was formerly fuppofed to contain 40,000 houfes and 240,000 inhabitants, but it muft be now much more confiderable; and Dr. Wilkinfon, an inhabitant of Effex, but who is a native of the county, and has feveral eftates in it, particularly Morley Hall, near Leigh, the place where the celebrated Leland took fome of his diftances, and who was a relation to a former poffeffor, a well-informed man, feemed to think that Lancafhire contained as many inhabitants as the county of Middlefex *, which he eftimated at about a million.

* " The idea of Lancafhire containing as many inhabitants as Middlefex, and which is there eftimated at a million, ought certainly to be qualified and corrected, as it can by no means be admitted by the Political Arithmetician, without the moft authentic and unequivocal proof; fer, fuppofing its

lion. In a circle of three miles around Tildesley, Thomas Johnson, Esq. informed the surveyor there were 10,000 weavers.

Though this estimation may be overcharged, still the population is great. The towns of Manchester and Liverpool, from the most authentic information, together contain 140,000 inhabitants. The roads from manufacturing towns are a continued street, house adjoining to house. From authenticated lists it appears that 22,000 men have been enlisted in the towns of Manchester and Salford only since the commencement of hostilities with France, and from the whole county of Lancaster not less than 27,000 have been enlisted in the space of eighteen months.—The Lancashire Fencibles have been raised since this account was published.

The work just published by Mr. Stockdale, under the title of *A Description of the Country from 30 to 40 Miles round Manchester*, affords various documents respecting the population of some of the most important districts of Lancashire; but any conclusion drawn from them, as to the whole county, must be in great measure conjectural. Dr. Aiken however has been so good as to draw up the following observations, stating the grounds on which such conjecture may be formed.

" Actual enumeration having but in few instances taken place within late years, the principal data for the purpose of calculation, are *bills of mortality*. The proportion which the articles in these bills bear to the number of people is a matter somewhat difficult to determine; but fortunately we have an unusually accurate standard in the bills of the parish of Eccles, in which, along with the annual returns of christenings, burials, and marriages, there is an annual enumeration of the families and individuals. From an average drawn from the comparison of these articles for several years, it appears, that

its two great towns, Liverpool and Manchester, to contain '75,000 each, its four other principal towns 50,000 amongst them, 50,000 more in its manufacturing parts, and 50,000 more in its remaining parishes, this would give only 300,000; nor will any probable data give a number bearing any considerable proportion to a million."—*W. Pitt, of Pendeford, Staffordshire.*

the

the chriftenings have been to the whole number of people as
1 to 26; the burials as 1 to 28½, and the proportion of perfons to a family, as 5.6 to 1. The much higher proportion
of this laft, than what has ufually been found in other places,
muft probably be owing to the great influx of children from
London and other parts to work in the cotton mills, who are
apprenticed and boarded with the inhabitants, and thus augment the number in each family. For the fame reafon the
deaths run higher than in country parifhes in general. The
article of chriftenings feems moft to be relied upon as a common ftandard of population; and it will probably be a calculation near the truth to multiply the regiftered chriftenings in
any town for a term of years by 25 or 26, in order to gain the
exifting number of people. In Eccles, the returned chriftenings are only thofe of the eftablifhment, but the return of families and people includes diffenters.

" Before we proceed, it is to be remarked, that from the year
1792, a very confiderable reduction appears in the bills for
almoft all the manufacturing towns ; but as this is owing to
caufes, it is hoped, merely temporary, particularly the abfence
of a great number of men in the army and navy, it would be
unfair to take the laft year or two as the exifting ftandard. I
have therefore, in the following calculations, made an average
of the chriftenings during the laft three years, in order to eftimate from them the actual population.

" To begin at Manchefter, the center of the moft populous
part of the county, and of the cotton manufactory. It's inhabitants, by the above rule, would amount to bout 63,000.
But by an actual enumeration in 1788, the *townfhips* of Manchefter and Salford were found to have only 50,000, and the
increafe of births fince that time, upon the average of the laft
three years, would only augment the number about 800. The
return of births muft therefore comprize a part of the *parifh*,
and yet only a part, fince at the enumeration in the year 1773,
the parifh was found to contain 13,786 inhabitants, and it may
be prefumed that the number is nearly doubled fince that period. On the whole, it will probably not be too much to

F f   fet

set down the population of the whole parish of Manchester

at — — — —	75,000
That of Eccles is about — —	14,000
Ashton under Line — — —	13,000
Prestwich — — —	6,600
Oldham — — —	17,000
Middleton — — —	6,000
Rochdale — — — —	15,500
Ratcliffe — — — —	2,000
Bolton — — — —	12,000
Bury — — — —	12,500
	173,600

" The above parishes are the whole, two inconsiderable ones excepted, in the hundred of Salford, which occupies all the south-eastern part of Lancashire, undoubtedly the most populous of its districts. If the number be raised to 180,000, it is supposed that all deficiencies in the calculation will be sufficiently provided for.

" The next hundred in size and population is that of West Derby, comprizing all the south-western part of the county, and containing the great port of Liverpool. This town, including all the new buildings within the limits of its township, probably contains about — — 60,000

" Of other parishes within this hundred, we have the following estimates :

Wigan — — — —	15,400
Leigh — — —	9,900
Warrington — — —	12,000
	97,300

" Though these are the most populous places, yet as there are many large and well peopled parishes, of which we have no account ; it will probably not exceed the truth to state the population of West Derby hundred at 140,000.

" Having thus made a rough estimate of all the southern part of Lancashire, the chief seat of its trade and opulence, the re-

remainder

mainder can only be the fubject of mere conjecture. Of the towns, we have documents to ftate

Prefton, at about	—	—	6,000
Chorley	—	—	4,200
Blackburn	—	—	12,100
Haflingden	—	—	5,400

There are no other towns of confequence, but

Kirkham, which may poffibly contain	—	5,000
and Lancafter	—	—	—	10,000

$$\overline{\qquad 42,700 \qquad}$$

" The remainder of the population of the county is divided over a large tract, generally thinly peopled, where trade and manufactures have not made their advances, as may be concluded from the fmall number of parifhes into which the county is divided. The tract called the Filde, between the Ribble and Wyer, is almoft entirely agricultural, and has the fcattered population ufual to fuch diftricts. The part bordering on Yorkfhire moftly confifts of wild uncultivated moors, fupporting a very thin population. The detached part acrofs the Lancafter fands is a rough and hilly region, little peopled, except in its lower grounds near the fea, and the neighbourhood of its mines. On the whole, if the number of 362,700 ftated in the preceding eftimates be raifed up to 425,000, by allowance for all the fmall towns and villages in thefe remote parts, it is fuppofed that the full population of this county will be given.

" One circumftance, however, ought to be mentioned, which may raife higher the idea of the population of Lancafhire in the minds of fome perfons. In the affeffment of men for the navy, laid by a late act of parliament on the feveral counties of the kingdom, and faid to be calculated according to the number of *rated houfes* in each, Lancafhire is placed higher than London and Middlefex together, the number for the firft being 589, and for the latter 552. Now, if this gives the true proportion of the *rated* houfes in each, that of the *unrated* muft probably be much larger in Lancafhire than in London and Middlefex, the rent of houfes being on an average much greater in the

F f 2   latter

latter than the former. It is not in my power to afcertain how the fact ftands in this particular, or whether any different rule was followed in the affeffment for London from that obferved in the country. But it is to be remarked, that the Borough of Southwark, and the parifhes within the bills of mortality on that fide the water, are not included in the above affeffment for London and Middlefex; and at any rate, it may be more juft to lower our notions of the population of the metropolis, than without due grounds to raife thofe of the population of Lancafhire."

Thus far Dr. Aikin, whofe fentiments upon the fubject are intitled to great weight. Mr. Yates, on the other hand, who had an opportunity, when drawing up his map, of minutely examining the ftate of the county (who is a man of keen obfervation, and lets few circumftances efcape him) is of opinion that the population is confiderably higher.

A gentleman calculates; that if Yates's map was divided into fquares, and the houfes in a certain number of fquares counted, and a medium taken, by allowing fo many perfons to each houfe, a tolerable eftimate might by this method be made. But Mr. Yates himfelf thinks fuch a medium would be much below the true ftate, fince from the fcale of the maps, many houfes and cottages were unavoidably omitted; befides, the number of people in each houfe of manufacturers, contains a greater number of inhabitants than are generally imagined, fome fmall buildings contain, it may be, two or more families, and the families not the leaft numerous.

If the clergy would afford their affiftance in fo important a bufinefs, (and there fcarcely remains a doubt but they would contribute their aid, if requefted, in circular letters directed to the rector or vicars of parifhes by the Board of Agriculture,) an eftimate might be obtained of the real ftate of population at a trifling expence.

CHAPTER XVI.

OBSTACLES TO IMPROVEMENT;

Including General Observations on Agricultural Legislation and Police.

THE obstacles to improvements are so many, that it is doubtful whether the whole can be here enumerated.

The grand obstacle is the want of a general inclosure act.

The great expence in obtaining particular acts, for certain districts; the odium, and ill-natured reflections, cast upon individuals who take an active part in promoting these good works, with the vexatious delays of frivolous obstructions, and many other causes, are obstacles of such magnitude, as to prevent even an attempt at an inclosure-bill, by the means of which many thousand acres of land, which lie waste and unprofitable, either to individuals or the public, might bear the richest grains, or fatten the choicest bullocks.

The corn laws have hitherto operated most essentially against improvements. If these matters were left to the simple operation of merchandise, and to find their own level by abundance, or deficiency, the farmer and the public would generally be benefited. Apprehensions of famine, under the present enterprising system of merchants, is entirely vanished. There will always be people bold enough to speculate in such an article of universal consumption, as to prevent a scarcity. The laws have hitherto afforded no assistance to the farmer. If there be a general failure of crops, the loss falls totally upon himself; he cannot avail himself of advancing the price, as a recompence for the failure of quantity*. The ports are opened for farmers or merchants to send in their produce from foreign nations,

* The question under consideration at present, is not what may most be conducive to the general good of the community, but what may be most advantageous to the farmer and fair trader. It is, in general, some adventurous speculator who reaps the most advantage, by artfully evading, or turning the law to his own favour.

whose

whofe lands pay no taxes to fupport our government, and fome of which are exempt from tythe laws.

Thank God that thefe laws have not hitherto wanted active oppofers, to whom the landed intereft lie under unfpeakable obligations. The averages, to govern the exportation and importation of corn, are formed from the mere declarations of *interefted* dealers, and cannot be juft grounds to regulate fo important a branch of commerce, which perhaps had beft be free, referving to the king in council, a power to interfere in cafes of great and fudden emergency. The expence of the corn returns throughout England, is very confiderable. In *Lancafhire*, a burden of near 600 *l. per annum* is fuftained for the falaries of corn-infpectors; although from the *corn act* it was fuppofed, the duties on *foreign corn* imported, were appropriated to pay all *thefe* falaries.

Tythes * are univerfally acknowledged to operate as obftacles to improvements; and they fall more heavily upon the fpirited agriculturift, than upon the indolent farmer. The greateft fervice the Board of Agriculture can perform to their country, will be to devife and carry into execution fome reafonable plan for their commutation.

The prohibition from exporting wool, in its raw ftate, is another obftacle againft encouraging the increafe of ftock, or paying that attention to the quality of fheep, fo as to produce the fineft wool; and fheep are reckoned the beft ftock for enriching either the arable or pafture farm. If liberty were given to export the raw material, under certain duties and reftrictions, the farmer would be benefited, the manufacturer would not be injured, and the revenue increafed.

The high duties upon falt operate as great obftacles to the application of this article to the advantage of their cattle, in certain cafes. It is an article moft cattle are fond of. It affifts digeftion; promotes a difpofition to fatten; prevents certain diforders; and, in foreign parts, they ufe it in large quantities, not being loaded by high duties. And, it is afferted, entirely

* Should not the incumbent of the day have a power to grant a leafe for 21 years certain, on fuppofition even of his dying the day after?

8

prevents

prevents that fatal difeafe among fheep, the rot *.—The refufe falt (an excellent manure) is thrown away, not being permitted to be ufed without paying the full duty ! ! !

Glebe, or church lands, or any other appropriated to the fupport of the meeting-houfes, and thofe lands which appertain to fmall livings, purchafed by the bounty of Queen Anne, are generally under a bad ftate of cultivation; the uncertainty of leafe, depending upon contingency of a fingle life, operating as ftrong obftacles to any degree of even moderate improvements; and in confequence they are, in general, under the very worft ftate of management.

Short leafes, moft certainly, are grand obftacles. The farmers would merit harfher epithets, than they are at prefent loaded with, were they to venture upon fpirited improvements for a fhort term.

Another obftacle to improvements is frequently occafioned by the obftinacy of an adjoining neighbour; *e. g.* one is difpofed to drain his lands, but cannot effect this without the concurrence of a fecond, or probably a third and fourth, to affift in fcouring ditches, opening water-courfes, and obftructions to the drains intended; and the difficulty of enforcing this concurrence, is, I fay, a great obftacle to many improvements. Where water proves injurious to roads, an opening may be effected, by application to juftices of the peace, and by indictment.—Why not admit of a fimilar operation, fo fimple and eafy to effect, in the practice of agriculture?

---

* It is to be lamented, that fome better method has not hitherto been devifed, to fecure the duties upon this article of falt, different from the expenfive mode of collecting it, by numerous officers; and, at the fame time, to take off the check given to the fifheries, and agriculture, by the high duties.

The money raifed upon the public, on the article of falt, in Great Britain, is £. 900,000, of which only one-third is received at the Exchequer.

The grofs revenue, in 1776, was	- -	£. 895,489
Drawbacks, bounties, and difcounts	£. 622,865	
Charge of management - -	26,410	
		649,275
	Neat produce - £. 246,214	

Vide *Knox's Tour*, p. cxlviii.

Vermin.

Vermin.—This is an object that requires more general attention than has hitherto been paid to it.

Individuals may have exerted themselves, and incurred great expence; but these exertions are of small avail, whilst furrounding neighbours are harbouring nurseries, to make future depredations upon those premises which they find untenanted. Several townships have, of late, associated together, and engaged a mole-catcher, at the rate of four-pence per acre, for a term of seven years; in which period of time the mole-catcher imagines he can nearly have destroyed the race of those animals in the district. This effort, towards a total extirpation, must be more efficacious than the greatest exertions of individuals. It is a doubt, after all, whether moles may not be useful animals in the destruction of certain noxious earrh-worms.

Rats are a very destructive animal, not only amongst grain, but other articles; they are frequently brought in abundance into the sea-ports in corn, and other vessels. The same mode has been very lately adopted, by particular townships, towards a general destruction of these very troublesome and voracious animals*.

Sparrows

* S I R,

　　" Through the vehicle of Mr. Young's useful Annals, I am informed of the establishment of a most excellent and honourable Board of Agriculture, under whom, I find, you are appointed to the survey of this county. To you therefore, I beg leave to address this, though it is not a direct answer to any of the queries proposed by the Board ; yet, I trust, it may be confidered, as having some relation to the former part of the last. This country is, to a very great degree, infested with that most destructive vermin, rats : I shall not, now, attempt any statement of the probable damages they may be supposed to do us; but the annual losses we sustain by them in our buildings, corn, and other goods, is very considerable. I, and most of the principal farmers, and others, for a circuit of about 20 or 30 miles, have, for some time, employed Edmund Heathcote, of Ormskirk, who has a very expeditious, effectual, and safe mode of destroying them ; but this affords us only a temporary relief, for we are, (perhaps from our neighbours, who had not theirs destroyed) before long, again infested.

　　" In some townships they have employed him to clear the whole for a stipulated sum, paid annually, out of some pound-rate-ley, which is so trifling, as not to be felt by any individual : and has, I hear, nearly the wished-for effect (*a*). But even this is certainly a plan too circumscribed to answer any great end. My reason, therefore, for troubling you with this,

　　(*a*) About one halfpenny in the affessed rates.

is,

Sparrows, and fmall birds, deftroy great quantities of corn; and fums of money have been annually paid, in this neigh-bourhood, towards their deftruction, for many years paft; and although the amount of the fum, from the number of years the cuftom has obtained, is become pretty large, no decifive effects have been produced; the premiums paid may have been too trifling to effect a total cure, and the meafures, hitherto taken, too languid. In this work, there ought to be an affociation, to declare war againft the common enemy; and vigorous exer-tions fhould be enforced, by fufficient premiums—for the de-ftruction occafioned by thefe fmall creatures is of greater ex-tent than many people could imagine. The amount of a hun-dred loads, facks of wheat, have been calculated to have been deftroyed by thefe diminutive devourers, in the courfe of one feafon, in a townfhip of no very large extent, befides the oats and barley. Magpies, carrion-crows, kites, hawks, and jays, fhould be included amongft the common enemy.

Dogs are in general a nuifance. The butcher frequently fuf-

is, in hopes, through you, to obtain, from the wifdom of the Honourable Board, fome fuggeftions for the moft eligible plan of extending the employ-ment of this perfon; or otherwife, for the extirpation of this moft deftructive peft. I am, S I R,

WIGAN, *in the County of* Your very humble Servant,
LANCASTER, Dec. 15, " OSKILL SUMNER."
1793.

The furveyor hath employed Mr. Edmund Heathcote, the perfon men-tioned in the letter, who always effected a prefent cure; but, after fome fpace of time, the vermin returned from other quarters. The man he be-lieves to be very fober and attentive to his bufinefs; pcffeffed of much civility, and has already obtained a certificate of his fuccefs, in places where the has been employed—a confiderable number of the gentlemen in the neighbourhood. *J. H.*

It is greatly to be lamented that Mr. Heathcote's method of deftroying rats and mice is not generally known and practifed; if it was, there would be a total extirpation of thofe obnoxious and deftructive animals, for in one night he totally deftroys them (where he is employed) be they ever fo numerous, as can be well attefted by hundreds in the neighbourhood of Onmfkirk, who have employed him.

The compofition he makes ufe of he puts in their holes or burrows, and from the very fmall quantity he ufes, it is aftonifhing it fhould have fuch an effect: it will keep good two years. A farmer recommends for the deftruction of rats, one ounce of pounded quick-lime to four ounces of tallow cake, to be beaten together and made into bails, and placed in their runs, which has cleared many buildings. But it has been proved by experience, that an ounce of aerated barytes finely powdered, mixed with the tallow, in place of lime, is more effectual.

G g

tains heavy loffes, in the deftruction or difperfion of his fheep, in the vicinity of great towns, by marauding dogs; and thofe who breed fheep frequently complain of their flocks being greatly annoyed by the yelping of curs, and who will fometimes wantonly encroach upon their borders. The paffenger is but too often attacked by their troublefome and vociferous falutations. They are certainly a fit object of taxation, if thofe of real ufe could be excepted.

Dogs are fo great a nuifance in many parts of this country, as totally to prevent all ideas of keeping fheep.—I wifh to Heaven we had a dog-tax.

Six perfons have *lately* died in the neighbourhood of Manchefter, from the bite of a *mad-dog*, and with dreadful fufferings; and twenty perfons, under the apprehenfion of being affected, were received into the Manchefter infirmary in *one week*.   *T. B. Bayley.*

Nothing can be more defireable for this populous county, than an univerfal tax upon dogs.   *Mr. Taylor.*

Weeds, efpecially thofe which bear winged feeds, as the thiftle, dandelion, &c. fhould be declared common enemies, and treated accordingly. It is to no purpofe that a neat farmer cleanfes his ground from fuch noxious enemies, if a lefs attentive neighbour permit them to flourifh in the adjoining premiffes; the winds will difperfe the floating emigrants over the well, as the ill-cultivated field, where they will take poffeffion, without the permiffion of the owner.

Another deftructive fpecies of vermin is a kind of fnail or flug, which, during the day-time in April and May lies under ground, devouring the roots of corn; in the evening comes out, and attacks the blade. Three or four may be found fometimes upon the fame plant, and this is the time that fhould be feized for their extirpation; by drawing a heavy roller over thefe lands whilft the enemy is at work, particularly in a moon-light night, they may be effectually deftroyed. By this ftep, a crop of corn may fometimes be preferved.

When the air is warm, and the atmofphere moift, the greateft flaughter may be made, the whole family being then abroad. They fkulk under ground on any approach of cold.

CHAPTER

Chapter XVII.

MISCELLANEOUS OBSERVATIONS.

Sect. i.—*Agricultural Societies,*

MANCHESTER SOCIETY.

THERE has been a fociety of agriculture eftablifhed at Manchefter, for a number of years, which is conducted with fpirit; and the feveral premiums offered annually, have been frequently claimed, and adjudged. A report is annually publifhed, with the premiums, which are offered for the enfuing year, and a lift of the perfons to whom they have been already adjudged, is made public; but they have not yet publifhed any volume of papers which they may have received on different fubjects; and of which they are in poffeffion. The furveyor, when at Manchefter, waited upon the fecretary, and examined thefe papers, with a view of collecting fomething that might be of fervice to him in this Report. The papers are many of them upon important fubjects.

The Rev. Mr. J. Stainbank, of Halton-hall, writes, " That the principal great towns, through the different counties, at leaft where they choofe to form themfelves into focieties, fhould be connected with the Board of Agriculture, as emanations from that great body, and be fupplied thence with books of inftructions, and other affiftance during their infant ftate; and that each fociety fhould adapt fuch a fyftem of premiums, as would be moft conducive for exciting a fpirit of agriculture in, and promoting the greateft poffible improvement of, its refpective diftrict."

Similar hints have been dropped by other correfpondents, but not fo fully explained.

Mr.

Mr. Ecclefton conceives, " that a fpirit for improvement might be excited amongft the farmers, by occafional tours, every three or five years, undertaken by a perfon appointed by the Board, whofe report fhould be printed, the names of the improvers and improvements to be inferted, with proper eulogiums for their induftry and ingenuity, in order to excite, by emulation, others to fimilar exertions."

The fame gentleman obferves. " The moft certain way to bring the cultivation of this ifle fpeedily to the utmoft degree of perfection would be to eftablifh a fchool or college where the elements of Agriculture, with its neceffary attendants, chymiftry, botany, &c. fhould be taught, and the moft approved principles of draining, floating, fencing, plowing, fowing in drill and broad-caft, the difference of manures afcertained, and their excellencies pointed out. Each operation to be fhewn the pupils in practice, on a farm eftablifhed for the purpofe, under the Board of Agriculture.

" Were fuch an eftablifhment in being, and properly attended to, moft men of fortune would wifh their fons to go through a courfe of the juft principles of a fcience the moft beneficial to mankind, which would give a turn of mind to the firft clafs of men in the kingdom, to encreafe its refources, by ameliorating their private fortunes, and greatly add to the comforts of the labouring clafs of people. The agents or ftewards of large eftates, who at prefent, from want of early inftruction, are unequal to their fituation, from the confined ideas of their education, would be able, along with the opulent farmers, to fend their fons with advantage, to receive all neceffary and folid inftruction, requifite for their line in life, befides *Arithmetic, Planning, and Surveying*, which at prefent is all that has been taught, even to the moft enlightened of that clafs, I may almoft fay of opponents (from want of better education) to modern improvements.

" The moft effential objects for the improvement of this county, are, the improved method of draining : the plafhing or making good fences : the introduction of green fallow crops, and the ftocking with fheep, for the fecurity of which ftock, in thefe populous parts, a dog tax would be highly
                                                                advantageous.

advantageous. All other improvements would of courſe follow."

OLDHAM SOCIETY.

There is a ſociety of botaniſts in Oldham, eſtabliſhed about twenty years ago, begun originally by Dr. Haulkyard, George Hyde, and John Newton.—The ſociety meets nine months in the year, and each member contributes ſix pence a month, (the preſent members are all artificers) two pence of which is reſerved for the purchaſe of books, and the remaining four pence ſpent in liquor.—They have purchaſed by this means about twenty volumes, and are poſſeſſed of 1,500 ſpecimens of plants, properly claſſed.

The time by many dedicated to paſtime, or ſometimes to worſe purpoſes, is by the members of this ſociety uſually employed in the purſuit of their favourite amuſement of either ſelecting or arranging their ſpecimens.

In collecting plants different members have gone as far as Liverpool, Lancaſter, Cheſter, Nottingham, Huil, &c. and one of the members has undertaken a voyage, and to proceed as far as the weſtern parts of America, to botanize, under the patronage of John Lee Philips, eſquire, of Mancheſter.— On the 21ſt of June, in the preſent year, one of the members being upon the mountains near Oldham, diſcovered for the firſt time the *uvá urſa*.

This ſociety is not unknown to Sir Joſeph Banks, Dr. Withering, and others, from whom they have been favoured by correſpondence of letters.—They are a wonderful and reſpectable ſociety for their perſeverance, ſobriety, and the great knowledge acquired in the purſuit of this ſtudy.

Their great ambition is to viſit the botanical gardens at London; for which purpoſe the ſum of five guineas, they think, would ſuffice : but alas ! that ſum is not to be found *!

---

* As a proof of the zeal of theſe induſtrious people, it may be mentioned, that upon Mr. Philips noticing to one of the members, that he had obſerved a certain rare plant whilſt riding on the northern coaſt of Liverpool, he immediately ſat out in ſearch of it, and brought it to Mr. Philips ; and the plant is now growing in his gardens at Mount-pleaſant.

Names

Names of the GRASSES moſt common in the neighbourhood of Oldham, given by two members of the
Botanical Society there :

1. Anthoxanthum, very common ⎫
2. Alopecurus, - - Dᵒ   ⎪
3. Dactylus, - - - Dᵒ   ⎪
4. Poa, - - - - Dᵒ   ⎪
5. Feſtuca, ~ - - Dᵒ   ⎪
6. Bromus, - - - Dᵒ   ⎬Hay graſs.
7. Avena, - - - Dᵒ   ⎪
8. Holcus, - - - Dᵒ   ⎪
                                  ⎪
Weight of crops is in general   ⎪
   Alopecurus, Poa, and Bromus.⎭

1. Aira, - very common ⎫
2. Agroſtis, - Dᵒ   ⎪
3. Secule, - not very common ⎬Paſture land.
4. Arundo, - Dᵒ   ⎪
5. Lolium, - very common ⎭

SECT. 2.—*Weights and Meaſures.*

THE difference of weights and meaſures in this county are
ſo many, that if they cannot with propriety be called obſtacles,
they may with truth be termed incumbrances to the general in-
tercourſe of buſineſs, and clear comprehenſion of what time an
under ſimilar terms, but with different ideas annexed to them,
according to the object.

The rod in Lancaſhire is of no leſs than ſix different lengths
in different parts of the county ; namely, the ſtatute or 5 ½ yards,
6, 6 ½, 7, 7 ½ , and eight yards, to the rod, pole, or perch *.

                                                              The

---

* To hazard a conjecture upon the etymology of the word, and the various lengths of the meaſure, the rod or pole got out of an adjoining foreſt,
was moſt probably the primitive meaſure, but without any certain ſtand-
ard. A ſtraight rod or pole, of 5½ yards long, preſented itſelf ; and this
ſerved to meaſure a certain diſtrict. Another rod, or pole of a different
length, preſented itſelf to a different meaſurer, and that became his ſtand-
                                                              ard

The meafures are equally variable.· At Lancafter a load of wheat, beans, and peafe, is four and a half bufhels (Winchef-ter); barley, fix Winchefter bufhels; oats, feven and a half Winchefter bufhels *.

N. B.—Wheat has been fold lately by the weight of 280 lb.

At Ulverftone, a load of wheat is 4¼ Winchefter bufhels; oats, fix Winchefter bufhels.

At Manchefter, a load of wheat is fixteen fcore; a load of oats nine Winchefter bufhels; a load of beans five Winchef-ter bufhels; a load of potatoes twelve fcore and twelve pounds, wafhed; unwafhed, thirteen fcore.

At Liverpool, the town's bufhel is 34¼ quarts for oats, bar-ley, and beans, making exactly 36 quarts Winchefter, or one-eighth more than a Winchefter bufhel; and by the cuftom of trade, one given in at every fcore, or twenty-one bufhels; of late wheat, barley, and oats have been fold by weight, but never yet beans: wheat 70 lb. to the bufhel, barley 60 lb. and oats 45 lb.; and probably this mode by weight is the faireft for both buyer and feller; for, befides the difficulty of getting a true ftandard bufhel or meafure, the dexterity of corn-meters is fuch, that it is afferted † they can gain either to the buyer or feller from 10 to 20 per cent. in different modes of meafurement; that 5 per cent. can be obtained by this practice by even bunglers in the bufinefs: this is an enormous profit, and the unfairnefs of fuch practices merits the fevereft reprehenfion ‡.

ard for another diftrict. Thefe rods, or poles, being fet apart for that purpofe, and ufed again when occafion called; and in time became the eftablifhed ftandard of the diftrict. Hence, *fall*, from the fall of the pole, which covered a certain length.

* A load, fo denominated, it fhould feem, from the horfe load, in a fack, the weight a horfe could conveniently carry on his back. Every kind of grain, &c. was conveyed this way till very lately. The load is the lighteft in the mountainous parts.

† By a confiderable corn-merchant.

‡ It is enacted by 31 Geo. III. that a Winchefter bufhel of corn fhould weigh as follows:

	lb.					lb.	
Wheat	57	avoirdupoife.	—	Wheat meal		56	Flour, 45 lbs. of which
Barley	49	-	-	-	- Flour	48	fhould be equal to a
Bigg -	42	-	-	-	- ditto	41	Winchefter bufhel,
Oats -	38	-	-	-	—	32	unground.
Rye -	55	-	-	-	—	53	

§   At

At Lancafter they have a meafure called a *windle*, which is three Winchefter bufhels.

At Prefton the windle of wheat, beans, and barley is three and a half Winchefter bufhels; but of late 220 lb. has been reckoned a windle of wheat; they have alfo a meafure at Prefton called a peck, which is twenty-eight quarts, four of which are called a windle.

Weights.—There are three different weights expreffed under the general term, *hundred weight*; namely, 100 lb. 112 lb. and 120 lb. The ftone varies. In Liverpool 20 lb. is the weight allowed for the feveral articles under that denomination, as beef, hay, ftraw, &c. and probably all the articles produced from land.

Butter is required to weigh 18 ounces, avoirdupoife, or may be feized by the magiftrates.

CON-

Conclusion;

*Means of promoting the Improvement of the County of Lancaster;
and Hints thence to be derived for the Improvement of other
Counties.*

A REPORT formed on so great a scale, as those which are
drawn up for the consideration of the Board of Agriculture, ought to conclude with a general view of those measures, which are best calculated for the improvement of the
district to which the survey relates; and also with a state of
those improvements which have taken place there, and by
adopting which, other districts might be benefitted.

I. *Considerations respecting the farther
Improvement of the County of Lancaster.*

In the preceding observations, a number of hints have been
given, pointing out the improvements of which this country
is capable; and it is only necessary to recapitulate some of the
most important.

1. *WASTE LANDS.*—The cultivation of the waste lands
in this county, is undoubtedly the first object that ought to be
attended to. A county like that of Lancaster, distinguished
for the opulence and spirit of its inhabitants, should never rest,
whilst a single acre remains, that does not yield some valuable
production. There is scarcely a rood in it, that might not
yield some species of grain, or some sort of useful pasture, or
some kind of valuable timber. Were those waste lands made

H h as

as productive as they ought to be, there would probably be no occasion for the importation of grain from other countries; and thus the manufacturing industry of Lancashire, instead of being a market to encourage the agricultural exertions of other countries, would be the means of promoting those domestic improvements, which, in every point of view, are so much entitled to be preferred.

2. *DRAINING.*—In a wet climate this must be the basis of all improvement. Much in this respect has been already done in Lancashire, but much still remains to be effected, particularly where the soil is of a clayey nature. The perfection however to which this art will probably be brought, in consequence of the attention which has been lately paid to it, and the discoveries which have been made by Mr. Elkington, will soon enable the people of this county, to clear their lands of superfluous water, whether it arises from what falls upon the surface, or is occasioned by subterraneous sources.

3. *GRAINS.*—Oats seem to be the natural grain to be extensively cultivated in this part of the island: and as in all countries an early species is desirable, it may not be unworthy of the Lancashire farmer, to try a species of oat that has lately been much cultivated in the neighbourhood of Edinburgh, known under the name of *the Red Oat.* It is remarkably early, being ripe before almost any other sort, and produces more meal than any oat of the same size; its straw also is good for cattle, and it is not liable to shake. It is probable, on the whole, that it is one of the greatest means of improvement that could be introduced into Lancashire.

4. *TURNIPS.*—An increased culture of this valuable root, is an object well entitled to the particular attention of those, who wish to promote the improvement of this county. A great part of the soil of Lancashire is supposed to be particularly well calculated for the culture of turnips. The advantages which other counties have reaped from this culture, ought to induce the Lancashire farmers, to pay particular at-

x   tention

tention to this fource of improvement, the nature and principles of which are too well known to require any elucidation
in this place. There are two modes of cultivating turnips;
the one is by the broad-caft, the other by the drill fyftem or
hufbandry. Which is the moft productive, has not yet been
decidedly afcertained; but the drill fyftem is the moft eafily
introduced, on the account of the greater facility of hoeing.—
For the broadcaft fyftem of turnip hufbandry, the furvey of
the county of Norfolk may be confulted ;—for the drill fyftem,
that of Northumberland.

5. *CATTLE.*—It is acknowledged that the Lancafhire
breed of cattle, do not equal what they were fome years ago,
and are certainly much inferior to the improved ftock of the
fame breed (namely, the long-horned) in other parts of the kingdom. As Lancafhire muft always be as much of a grazing,
than of an arable country, it is particularly defirable, for the
advantage of its inhabitants, that the herbage it produces, fhould
feed as profitable a fpecies of ftock as poffible; and hence particular attention to its breed of cattle cannot be too ftrongly
recommended.

6. *SHEEP.*—It is impoffible to fee without regret, that fo
valuable an animal, fhould hitherto have had fo moderate a fhare
of the attention of the Lancafhire farmers, as there is none by
means of which fuch great improvements might be effected.
Notwithftanding the humidity of the climate, where the foil is
dry, or capable of being drained, no apprehenfion need be en-
tertained of this animal's fucceeding to a wifh. At prefent, the
greater part of the county feems to be principally devoted to
the moft unprofitable of all that fpecies of ftock, namely, the
black-faced Scotch, whofe fleece is of little or no value, whofe
reftleffnefs renders it difficult for them to be confined in any
common inclofure, and the wildnefs of whofe difpofition makes
it extremely difficult to fatten them. Inftead of thefe, there are
two forts of fheep, the Cheviot for the hilly parts of the country,
and the Bakewell or Culley breed, for the lower diftrict, which
cannot be too ftrongly recommended to the people of Lancafhire.
The Cheviot are to be found on the borders of England

H h 2   and

and Scotland, and are the moſt valuable breed, for a mountainous diſtrict, perhaps any where to be met with; but for a
manufacturing country, where the paſture is ſufficiently rich,
the Bakewell breed is undoubtedly preferable to every other;
producing, from the ſame extent of herbage, a greater quantity
of meat, and of a ſort peculiarly well calculated for general
conſumption. In a manufacturing diſtrict alſo, it is extremely
deſirable, to have a raw material of ſuch value as wool, on
which the induſtry of the people may be exerciſed, ſhould
other branches fall off.

Theſe General Obſervations might be extended to a much
greater length, and might include a number of other particulars: but if the *waſte lands* of the county are properly cultivated—if *draining* is properly attended to—if the beſt ſpecies
of *oats* and other grains are propagated—if the culture of *turnips* is carried to that extent of which it is capable—if the
cattle of the country are improved, and regain their ancient
eſtimation—and, above all, if the beſt ſorts of *ſheep* are ſpread
over the county, Lancaſhire will have no reaſon to regret
the attention that has been paid to its improvement by the
Board of Agriculture.

II. *Hints for the Improvement of other Counties.*

THE attention of the people of Lancaſhire, has hitherto
been principally devoted to the extenſion of manufactories;
at the ſame time, an active and intelligent race of people, muſt
always diſcover a number of particulars, by which its own
agriculture, and that of its neighbours, may be improved. A
variety of hints to that effect, will be found in the preceding
pages of this Report; but there is one point which requires to
be particularly adverted to, namely, the management of marle,
in which this county ſeems to excel every other, and by imi-
tating whoſe practice, there is no part of the kingdom, where

marle

marle might be found, that might not be brought into a high ftate of cultivation. The quantity laid upon an acre feems very great, but is amply repaid by the lafting benefit that refults from it, It is probable indeed, that a fmall quantity may do little good, whilft a great load may produce the moft important benefits. The marling alfo a fecond time with great advantage, is a circumftance entitled to very particular attention ; and the burning of marle, and ufing it when burnt as top drefling for corn, is a mode of improvement which cannot be too ftrongly recommended to the attention of the induftrious farmer, who has an opportunity of putting it in practice.

On the whole, it is believed, that no man can read over the preceding pages, without being fatisfied, that great pains muft have been beftowed in collecting and arranging fuch a mafs of valuable information ; and if a fimilar account is drawn up and printed of every other diftrict in the kingdom, there can be no doubt of its proving in the higheft degree ferviceable to the country.

APPENDIX.

APPENDIX.

Nº 1.

THE following defcription of the Lancafhire cattle, &c. will ferve to explain the engravings which accompany this Report.

LANCASHIRE BULL,

Was bred at St. Michael's in the Filde, and is now the property of Edward Afhcroft, Spellow-houfe farm, in Walton.

DIMENSIONS.

	Feet.	Inches.
Length of the head - -	1	8
Depth from fhoulder to breaft-bone	2	7
Breadth from hip to hip - - -	2	6
Height from fhoulder to fore-foot -	4	7
Length from root of horn to rump -	8	1

LANCASHIRE COW,

Purchafed, when in the poffeffion of James Balmer, Toxteth Park, for exportation to America, as one of the beft fpecimens of the Lancafhire breed.

DIMENSIONS.

	Feet.	Inches.
Length of the head - -	1	4
Depth from fhoulder to breaft-bone	2	3
Breadth from hip-bone to hip-bone	1	11
Heighth from fhoulder to fore-foot -	4	2
Length from root of the horn to rump	7	4

LANCASHIRE MARE,

Bred at Weft Derby; is of the ufual breed of cart-horfes in that vicinity, ftrong and bony; the colour black, not fo heavy but that it might occafionally be ufed upon the road, or to draw in a chaife.

The

The Mare from which the original drawing was taken, is in her 22d year, notwithſtanding which the teeth are yet good, eyes clear, and perfectly found. It has been one of the beſt of ſervants, to its preſent maſter, for the ſpace of nineteen years.

MIXT BREED OF HOGS.

The Hog, an engraving of which is inſerted in this work, is a boar belonging to Thomas Wakefield, Eſq. Brooke Farm, near Liverpool. There is a mixture of the Chineſe and of the wild boar in this breed. Its chief properties are a large carcaſe, ſhort legs, ſmall entrails, and great weight of meat, in proportion to its ſize.

DIMENSIONS.

	Feet.	Inches.
Length of head - - -	1	0
Depth from ſhoulder to breaſt -	1	4
Breadth from hip to hip - -	1	0
Heighth from ſhoulder to fore-foot	2	6
Length from ear to rump -	4	0
Girth round his body - -	5	2

N° 2.

Mode of preſerving CREAM, for ſeveral weeks or months; particularly calculated for ſea voyages.

TAKE 12 ounces of white ſugar, and diſſolve it in ſome ounces of water, over a moderate fire. After the ſugar is diſſolved, boil it for about two minutes in an earthen veſſel; after which add immediately 12 ounces of freſh cream, and mix the whole uniformly over the fire: then ſuffer it to cool, pour it into a quart bottle, and cork it carefully. Keep it in a cool place, and it will continue fit for uſe for ſeveral weeks, or even months.

N° 3.

N° 3.

SINCE the above Report was drawn up, the following paper was tranſmitted, by an active and intelligent magiſtrate for the county of Lancaſter.

Obſervations on the Corn Act, 31ſt Geo. III. chap. xxx. reſpecting the Salaries of the Corn Inſpectors.

It is inſiſted, that the clear meaning of the legiſlature was, to defray the expences of its execution, and amongſt theſe the ſalaries of the corn-inſpectors, from the duties to be paid on the importation of foreign corn.

In proof of this—

1.

Be it obſerved, that by the 15th, 16th, 17th, and 18th clauſes, various duties are impoſed on foreign corn imported; and are put under the management of the commiſſioners of the cuſtoms.

2. That by the 74th clauſe, and the two following ones, expreſs proviſion is made for the *re-payment* of the monies paid by the county treaſurers, viz. (5 *s.* for each return) charging *alſo* the deficiency (if any) to the general cuſtom-houſe account to make good.

3. That the regulations for the port of *London*, in clauſes 43, 44, 45, and 46, provide for the ſalary of the corn-inſpector *there*, from the duty of one penny on *Britiſh* corn, and two pence for foreign corn imported.

This is plainly done from the juſt view of the ſubject—as of *national*, and not of *local* concern; and therefore no partial burden is thrown on the city of *London* to pay their corn-inſpector; and there can be no doubt, that on the ſame principle of equity, all other parts of the kingdom were intended to be equally exempted from *local* impoſitions.

4. That

4. That the reafon of the allowance made to the Scotch counties (by the 33 Geo. III. c. 65. fect. 20.) is declared to be, that the former allowance of twenty fhillings for each return (by 31 Geo. III. c. 30. fect. 74.) was not fufficient to defray the expences, &c. This fully explains the meaning of the legiflature in the corn act, *not to* burden the particular " counties" by the payment of extra falaries, &c. The act of 33 Geo. III. c. 65. puts it out of all queftion, with refpect to the counties in *North Britain* ; and as both parts of the united kingdom are under the regulation of this corn act, the fame meafure of equity *muft* apply to both.

5. It was calculated when the act paffed, that the duties on foreign corn imported would be more than fufficient to defray the expences of the act ; for the " furplus" is ordered to be paid to the receiver-general of the cuftoms. And the fums *actually* remitted *on this account* from Liverpool, will prove that there is no neceffity (if that were to be admitted as a plea) to burden the county rates of " Lancafhire" with the payment of £. 500 per annum for the falaries of the corn-infpectors within *that county*.

6. If it was judged proper to order the *fmall payments* " of *five* fhillings" for each return to be *repaid* to the counties, it muft follow that the legiflature never meant *locally* to burden, and to fo great an extent, any diftricts within the united kingdom, to fupport a fyftem of *general regulation* ; and for which adequate provifion was intended to be made in the corn act, by the fmall duties laid on foreign corn imported ; and which in fact, are fufficient for this purpofe.

Hope, near Manchefter,
    April 1795.

                                T. B. BAYLEY.

                THE END

                    I i

Directions to the Binder.

For EU product safety concerns, contact us at Calle de José Abascal, 56–1°, 28003 Madrid, Spain or eugpsr@cambridge.org.